THE
SPITFIRE KIDS

The generation who built, supported and flew
Britain's most beloved fighter

ALASDAIR CROSS
with David Key

HEADLINE

First published in the UK in 2021
by HEADLINE PUBLISHING GROUP

1

Cataloguing in Publication Data is available from the British Library

Hardback ISBN 978 1 4722 8196 8

Designed and typeset by EM&EN
Printed and bound in Great Britain by Clays Ltd, Elcograf S.p.A.

Headline's policy is to use papers that are natural, renewable and
recyclable products and made from wood grown in sustainable forests.
The logging and manufacturing processes are expected to conform
to the environmental regulations of the country of origin.

HEADLINE PUBLISHING GROUP
An Hachette UK Company
Carmelite House
50 Victoria Embankment
London
EC4Y 0DZ

www.headline.co.uk
www.hachette.co.uk

THE
SPITFIRE KIDS

Contents

Introduction

Southampton doesn't care much for the land. Of course it has some lovely old buildings and leafy suburbs, like most small English cities, but that's never where your eye is drawn. It's a city built from water. Stay a while and it's the land that seems transitory, ebbing and flowing while the Itchen, the Solent, the Channel and the Atlantic are the city's permanent mooring points.

The seaplanes that launched from the Woolston slipway, the warships built at John I. Thornycroft next door, and the *Titanic* and all the other great transatlantic liners that slid from their berths across the river: they're what define the city, all of them straining to escape to the ocean, all drawn to the great world beyond.

In 1940 this town, so used to looking out, was suddenly forced to retreat inward. Its people looked to the surrounding fields and woodland for shelter from incessant bombs. They looked deep within for the reserves of courage they would need to rebuild the smouldering wrecks of their docks, factories and homes.

*

Geoff Wheeler, my guide to Southampton's riverside, is a brilliant spinner of tales. His uncles worked in the Spitfire factory and his father was a policeman who endured shift after shift without rest at the height of the Southampton Blitz. With a wave of his hands and a few well-chosen words, these deathly quiet docks and warehouses are once more buzzing with industry, the river alive with grand ships. And in the sky? Spitfires, of course.

Geoff unlocks a gate in a chain-link fence and within a few steps we're on a greasy black-brick slipway. This is the last relic of the original Spitfire factory. It's here that workers stopped for lunch, dipping their tired feet in the River Itchen. It's here that they sheltered from German bombs, caught out in the open because the siren sounded too late or not at all. And it's here that elegant silver sea-planes first slipped into the water – the planes that would become the Spitfire.

In the 1930s, Geoff tells me, this was the Silicon Valley of its time, a true hi-tech hub. It was here that the world's fastest planes were built alongside cutting-edge destroyers and corvettes, and the vital parts for almost every aspect of mechanised modern life. Geoff pulls out a sepia photo of the Supermarine Woolston works taken from the air. Right where we're standing, by an enormous concrete pillar supporting the thundering traffic of the Itchen Bridge above, this was K Shop, where the very latest in aero technology was bent and shaped for flight.

Southampton and Portsmouth, its neighbour and fierce rival, formed a large and thriving jobs market. Young apprentices would move from factory to dockyard in search of the best wages. The workforce was motivated, highly adaptable and familiar with all the latest manu-facturing technology. There really was nowhere better to design and build the Spitfire.

Perhaps it was those twin aspects of the place – its modernity and its instinctive need to lean outward to the world beyond – that made the Luftwaffe raids on the Spitfire factory and the Blitz that followed so traumatic. A city dedicated to making the future was smashed from the air by the latest technology of a merciless enemy.

Since the end of the First World War, politicians, strategists, novelists and film-makers had been scaring the public with the idea of death from the air. Bombers were unstoppable, cities could not be defended. The next major war would be won by the nation with the biggest bomber force. It would pound cities into submission until the people begged their leaders to surrender.

The 1936 science fiction film, *Things to Come*, written by H. G. Wells and produced by Alexander Korda, imagined a war breaking out in December 1940. A small British city is attacked by a fleet of bombers. Anti-aircraft guns fire desperately, cinemas artfully collapse, a gas main bursts and flames envelop the city. Within minutes there is nothing left but dust and death.

The mood of existential dread was darkened further by newsreel footage of the Condor Legion's raid on the Basque town of Guernica in April 1937. The sheer destructive power of the German bombers was captured in *Guernica*, Pablo Picasso's black-and-white vision of chaos and suffering. His modernist masterpiece toured Europe, shocking audiences in London, Leeds, Liverpool and Manchester. A 1937 report from the government's Committee of Imperial Defence estimated that a sixty-day aerial bombardment of Britain would kill 600,000, wound 1.2 million and cause a deep psychological impact out of proportion to even these enormous figures.

In the summer of 1940, the people of Southampton could take pride in the fact that these terrible visions had yet to become a reality in Britain. German bombers were mounting mass raids on targets in southern England, but day after day they were challenged by the beautiful, brilliant fighter built on the banks of the Itchen. As long as the workers of Southampton could keep turning out Spitfires, then the Luftwaffe could surely be kept at bay.

*

When we first began planning our BBC World Service series, *Spitfire: The People's Plane*, it was those people that really sparked my interest – the young men and women chucking their lunchtime crusts to the seagulls one moment, throwing themselves to the ground the next. How did they feel, finding themselves abruptly shoved to the front line of a world war? Many were teenagers – factory apprentices or office workers enjoying their first pay packets. They had barely tasted adult life when they huddled behind this slimy slipway, weighing up their chances in the face of death. Should they wait for a lull and make a run for the shelters or stay here, hidden from the strafing machine guns but still vulnerable to the blast of bombs?

Eighty years on, the myth of the stiff upper lip – a useful fiction in the face of overwhelming odds – has ossified into accepted fact. Our collective memory of the Battle of Britain is permanently fixed, populated by a cast of handsome, upper-class pilots and plucky workers always ready for a chorus of 'Roll Out The Barrel'. It doesn't leave much room for emotion – for love and hate, boredom or fear. It means that we miss the crucial fact that so many of these Home Front heroes were in their teens and early twenties. Whether they were panel-beaters, draughtswomen or fighter pilots, they were raw and vulnerable. Like most teenagers, they felt every emotion keenly. The highs were higher and the lows lower.

Imagine the ground tremble and kick as bombs drop around you, feel the adrenalin surge as your plane twists and turns to escape the tracer-spitting Messerschmitt on your tail. Add in the confusion of hormones and the tangled emotions of any teenager and you have an intensity of experience that's almost too much to process.

In writing this book and expanding on the stories we told in *Spitfire: The People's Plane,* I want to render those experiences raw once more. Read their diaries or listen to interviews with those who played a part in the Spitfire story and you'll certainly find plenty of stiff upper lips, but listen again, read a little deeper, and the teenagers beneath reveal themselves; frightened, confused and astonishingly brave.

Writing during the First World War, the French novelist Marcel Proust described how most of us struggle to pinpoint the vital moments that formed our personality. The passing of the years conjure an illusion that our 'self' has been created incrementally, by a gradual drip-feed of experience. For the young people in this book, there is no doubt which period of their life transformed their souls, truly made them as men and women or – in so many cases – broke them.

*

In 2008 the Spitfire pilot Geoffrey Wellum took the BBC's Clare Balding for his favourite walk around the Lizard peninsula in Cornwall. Geoffrey realised something important during the Battle of Britain, and he was able to confirm it sixty-eight years later. Flying as a teenage pilot in a Spitfire squadron in 1940 was the pinnacle of his life. Nothing since could begin to compare with the heights of ecstasy and depths of depression he experienced back then. He told Clare that medals and memorial fly-pasts didn't interest him:

> Nobody wants a thank you, nobody wants a medal. I think the thing that we want is to be remembered, because that takes in all the people who were shot down and didn't get a medal and deserved it. We're not heroes. Don't give us medals. Just remember us. That's all.[1]

In the next few hundred pages you'll meet the young men and women who built, designed and flew the astonishing aircraft that turned the tide of war. They deserve to be remembered.

Part One

THE BUILDERS

1

The Perfect Target

It's too loud to chat. The typewriters, dozens of them, rattle and ping. The bass rumble of machinery four floors below flattened every few minutes by the heart-stopping thud of a drop hammer or the shrill shriek of a band saw. Metal against metal.

Just as well, maybe. It's not as though there's time to talk to the other girls. The paperwork's relentless. More and more parts coming in, more and more orders going out. The factory is running day and night, seven days a week, shifts have been increased to seventy hours a week. The factory and every worker in it close to breaking point.

The command from the Ministry is clear: 'More Spitfires, More Spitfires!'

It's August 1940 and Joan Tagg has one of the most important jobs in the war. Just turned twenty, she's a shorthand typist in the Wages Office at the Supermarine factory in Southampton, home of the best fighter plane in the world, the Spitfire.

Most of my work consisted of typing documents relating to the specification of the Spitfire aircraft supplied to RAF stations. I particularly disliked typing statements for the aircraft and parts. Each form had to have six copies, which meant sheets and sheets of carbon paper.[1]

Joan left Itchen Secondary School in 1938. She wants to train to be a primary school teacher, but for the moment her family need her to bring home a steady wage. Supermarine offered 25 shillings a week and – another big attraction for a keen athlete – a sports club where she could play tennis and netball. At the weekend she joins the young workers pouring into Southampton's cinemas or dressed up for a night of dancing with Brylcreemed boys at the Guildhall.

Jobs are plentiful in Southampton's factories and docks, as Britain urgently re-arms. The dark days of the Great Depression of the 1930s are gone but, by the summer of 1940, something infinitely more sinister lurks just over the horizon.

Britain stands alone against the Nazis. Hitler's forces have swept west, through Holland, Belgium and France. Their guns are close enough to shell the English coast. Invasion is imminent, but before the German Army can cross the Channel they need to wipe out the Royal Air Force. They've been doing a pretty good job of that, destroying airfields and aircraft on the ground and in the air. By August there's virtually nothing in reserve. The RAF squadrons desperately need new aircraft, particularly the Spitfire, the fastest and deadliest fighter plane of the war.

That makes Joan and her friends the number one target for the most powerful air force in the world, the Luftwaffe. And that makes for a long hot summer of sleepless nights:

I must admit that I'm getting rather jittery, being constantly on the alert and am also feeling the effects of disturbed nights. Even though planes may not be overhead we do hear gunfire and are constantly up and down to the shelter.[2]

Four floors below Joan is the source of that low rumble and all that shrieking metal, the factory floor. This is the Woolston works, one of two factories set close together on the banks of the broad River Itchen – Southampton's industrial waterfront. On the ground floor there's E shop, where the planes are assembled; P shop, where the wings are made; and K shop, where twenty-year-old Cyril Russell beats the panels that become the Spitfires. He's been at Supermarine for five years but vividly remembers the intense sights, sounds and smells of his first day:

> The wonderment of seeing and being close to aero-planes in their varied assembly stages! We passed through the Machine Shop with its whine, through the comparative calm of the Wing Shop and then into a gangway to a small wooden and glass office in the middle of a sea of benches, with men banging and hammering pieces of silver coloured metal.[3]

Cyril, a tall and skinny lad with a broad grin, loves every minute of his life at Supermarine. He enjoys the hustle and bustle and the blokey humour of the factory floor – all 'left-handed' screwdrivers, empty threats and cheerful insults – but most of all, he loves the aeroplanes. On his first day at work as a fourteen-year-old 'handy lad', he spent his lunch hour running his eyes and hands over a Seagull flying boat. The detail of the construction enchanted him. Seeing how metal and fabric could be somehow woven together to produce something so elegant set Cyril on his course for life. By the summer of 1940 Cyril is with a panel-beating squad, bending the crucial panels for the Spitfire fuselage.

Away from the factory, Cyril sees Joan every Sunday at the Methodist Church. And Cyril carries a torch for Joan's best friend, Peggy:

> I had quite a crush on her, and we had gone to the Grand Theatre together. She was a secretary and known to the lads as 'the girl in green' because of the smart green outfit she wore, with a little fur hat. Her name was Peggy. Peggy Moon from Canada Road and a lovelier girl one could not wish to know.[4]

Peggy Moon from Canada Road and her friend Joan Tagg had gone to school together. Joan left a year earlier than Peggy, but they're together again as typists on the fourth floor of the factory's office block. Joan keeps a diary. Until now it's mostly lingered on film stars, friendships and crushes, but in August and September 1940, she captures the feelings of the workers on the new front line, as Britain comes perilously close to losing the Second World War:

> We are very busy at work and it is becoming difficult to complete everything because of these constant interruptions for air raid alarms which mean frequent dashes to the shelter which is beside the railway bank in Sea Road.[5]

The air raid shelters at Supermarine are 500 metres from Joan and Peggy's office. To reach them, dozens of typists have to squeeze down four flights of narrow stairs

from the top floor. That leads them straight into the flow
of hundreds of factory floor workers making their own
way to the shelters. Once out of the Hazel Road exit, they
have to turn up the hill on Sea Road, cross under the railway bridge and sprint across the open expanse of Peartree
Green. Only then do they reach the shelters, half-buried
brick buildings squeezed into the only space free between
the railway line, the factories and the shipyard next door.

At first the air-raid warnings are a welcome break
from work – they're false alarms or a panicked response
to German reconnaissance planes high overhead, but Joan
soon gets a frightening sense of the true destructive power
of the Luftwaffe.

> Tuesday, 13th August 1940: We had a real air raid this
> afternoon. We reached the shelters and these were
> shaking as the planes came over, machine gunning
> and bombing – as we crouched inside they seemed to
> be swooping low over the shelter. We played the comb
> and sang to cover the noise – our favourites in these
> situations, 'Roll out the Barrel' and 'Bless 'Em All'.[6]

The Supermarine factory is vulnerable. Southampton
is a compact industrial city, turning out vital material for
the war effort. There's the Thornycroft shipyard building corvettes and destroyers for the Navy and the Pirelli
factory making cables, pipes and machined parts for all
sorts of planes, boats and munitions. As a target, the city
is irresistible and frighteningly easy to hit. From their new
bases on the French coast, the Luftwaffe can roar across
the English Channel in minutes. They can tuck into the
coastline of the Isle of Wight before gaining height and
turning toward their targets. It's impossible for the Royal

Air Force to guess those targets until it's too late. Will they head for the radar station at Portland or the coastal airfields? Will they fly north to the industrial Midlands? East to Portsmouth, home of the Royal Navy? Or west to Southampton?

Even if RAF controllers guess correctly and dispatch the Hurricanes and Spitfires to intercept the bomber squadrons, there's another fundamental problem of geography – Southampton is comfortably within the flying range of Germany's best fighters. With small fuel tanks, the fearsome Messerschmitt 109s have a strict limit on their time in the air over Britain. They can't stay with the bombers on raids to the north or the Midlands, but Southampton is close enough for the fighters to keep the Hurricanes and Spitfires busy as the bombers approach their targets.

The factory site has its own fatal flaws too. Phil Pearce, an eighteen-year-old apprentice at Supermarine, spent much of his childhood watching ocean liners and huge freighters come and go up the River Itchen. One particular ship caught his eye:

> I can remember a ship that came into the docks from Germany called the *Bremen*. A seaplane bringing the mail from abroad would be catapulted from this ship. It used to tie up at the jetty by the factory for refuelling and sometimes to be serviced. So, the Germans were well aware of what was produced in the factory and its specific location.[7]

The main factory building itself, meanwhile, couldn't have been more obvious. Even a last-minute coat of camouflage paint couldn't disguise the broad, white Art Deco frontage prominently positioned on the waterfront. And the media hadn't been shy in telling the world about the exciting new British fighter, with a newsreel crew touring the factory and the assembly hangars in nearby Eastleigh just months before war was declared.

Despite the vulnerability of the factories, nothing is done to improve the shelters or bring them closer to Cyril and Phil on the factory floor or Joan and Peggy in the office. Air raid sirens sound every day, workers scurry for shelter and production stops. As the Battle of Britain rages over southern England, fighters are shot from the sky and the Ministry demands more Spitfires from the Southampton works. The government's War Cabinet is worried. Too much production time is being lost with the workers sitting in shelters. Supermarine's management respond to the concerns with a frightening announcement. The horrors of global war edge a little closer to Joan and her friends:

> Friday, 23rd August: We have now heard that because so much time is being lost at work we are only to get the warning to leave when danger is imminent! This means that we will have even more perilous gallops to the shelters having come down several flights of stairs from the fourth floor.[8]

Instructions are issued for the factories to operate a 'delayed warning system'. This means that public air-raid alarms will be ignored and works sirens sounded only when spotters on the roof see enemy aircraft approaching.

That gives workers hardly any time to reach the shelters. It's early September now and there are more air raids on Southampton. Factories and dockyards close by are hit by bombs, friends and neighbours are killed and injured. For Joan the tension is unbearable:

> Monday, 16th September. Once more the siren sounded and we rushed to the shelter. Unfortunately we had to go the long way round as a bomb had fallen on the path to the shelters. I got a tin can caught on my heel so I was slowed up when climbing the bank. I am not looking forward to going to work tomorrow. We are so busy and I feel that we are likely to be attacked very shortly. I wish they would evacuate us to a safer place.[9]

Joan's friend Peggy is transferred to the Itchen Works, the other half of the Spitfire factory. It's closer to the shelters, but no one feels safe.

The management at Supermarine and the new Ministry for Aircraft Production aren't blind to the vulnerability of the Spitfire factory. At Castle Bromwich, 130 miles to the north, a massive new factory is being built to make use of the engineering skills of the area's carmakers, but the team in the Midlands is far from ready. If the Spitfires being shot down right now over southern England are to be replaced, then they can only come from the Southampton plants. The safety of the workforce is not the priority; Spitfires must be built right here until the Luftwaffe makes that impossible. Joan and her friends are about to face the consequences.

> Tuesday, 24th September. In the afternoon without warning there was terrific gunfire and we rushed to the window (how foolish).[10]

No sirens wail. The 'immediate danger' hooter is silent:

We heard the sound of planes and the whine of bombs and saw all the staff fall to the floor underneath the desks – bosses and all. Found myself laughing. We tried to make our way to the ground floor but the stairs were blocked with those too frightened to venture forth. We finally got to the basement, where we sat on boxes, only to be told by the soldiers on guard – 'I wouldn't sit there love, those boxes are full of live ammunition'.[11]

At Ludlow Primary School, a ten-minute walk from the factory, the children of the workers in the docks and aircraft factories fall to the floor and tuck themselves beneath their desks, just as they've been taught. There was no warning, just a terrible roaring as the noise of the anti-aircraft guns and the bombs merged into one horrific wall of sound. The headmistress, Miss Quayle, pushes Rowntree's Fruit Gums into their mouths with a reassuring cry of 'Don't worry girls, it can't last for ever'.

Back at the Woolston factory, Albert Hatch is engrossed in his work, busy with the final assembly of a Spitfire fuselage, completely unaware of his mates running for their lives:

First warning I heard was a gun going off and I was up in the tail of a Spitfire. Of course I came out of there like a jackrabbit. And made my way to the shelters and that was when all hell was let loose.[12]

Cyril Russell sprints from K Shop, but it's too late to reach shelter:

I looked up and to my left, there they were, with the bombs already leaving their racks. I flattened myself into the gutter. The exploding bombs made the ground tremble and kick, and for a moment or so, although it seemed longer, there was just nothing I could do.[13]

Phil Pearce joins the frantic rush for cover:

I got halfway down the stairs and saw a very large press which I felt would give me more protection so I quickly crawled under it.[14]

There's a lull in the raid and Phil makes a run for it, but before he reaches cover the second wave of bombers start their attack:

We were now out in the open – we quickly searched for some form of shelter and we could only find a small section of beach alongside the river . . . so we laid down there. I heard an aircraft and looked up and saw a German bomber flying very low over the river, the next thing I knew was the shingle on the beach flying up around me and I could see the rear gunner as he was strafing the area.[15]

Joan unfolds herself from the floor, brushes down her skirt and walks toward the shelter, but her route is blocked. Smoke rises from the shelters and the narrow passage way from the Itchen works to the shelters has been hit. She stands and stares in horror as the stretcher bearers stagger by with the dead and injured.

The factories got off lightly. The bombs have missed their target, but across the town forty-two people are dead and sixty-three seriously injured – many buried in the shelters or caught in the open as they ran for cover.

As she turns to walk away, Joan is given some devastating news:

> My friend Peggy had been one of those killed sheltering beneath the railway bridge. We had on the previous Saturday attended a friend's wedding and I walked home with Peggy talking about our favourite songs. Peggy's favourite was 'Begin the Beguine' . . . I felt very trembly and could not believe it had all happened.[16]

Cyril too is devastated:

> My foreman came over to me and told me quietly that one girl had been killed. It was the 'Girl in Green'. Peggy. Peggy Moon.[17]

Cyril and Joan know that Peggy's younger sister contracted meningitis just a few weeks earlier. She had been evacuated from the city and died far from her home and family. Another death, so soon, would crush the Moon family. For the Supermarine workers, though, mourning will have to wait. As the smoke clears and fire engines rush to the flattened old fishermen's cottages behind the factory, the workers are ushered back inside. Joan sits at her typewriter, certain of one thing:

> I will not use these shelters again – there are not many standing now anyway – we are all planning different hideouts nearer at hand. Some say they are going to run as fast as they can well away from the factory.[18]

By Thursday, 26 September, two days after the raid that killed Peggy Moon, Spitfire production is up and running again. There are empty spaces at typewriters and workbenches, but conversation turns to the day ahead, not the dead crushed in their shelters. Spitfire parts are being dipped, doped, bent, rolled and assembled on the factory floor and Joan is busy in the office:

> Peaceful morning – managed to catch up on my pile of invoices. We went across the road to the canteen for our tea and I was just raising the cup to my lips when the immediate danger hooter blasted out. With one accord we all rose and rapidly made our way to the exit – like ants from an anthill. We had no intention of trying to reach the shelters. [19]

Joan runs back to the office block and takes cover in the basement along with most of the Supermarine management team. This time there would be no mistakes from the German bombers – sixty Heinkel 111 bombers attack in two formations. Close protection comes from sixty Messerschmitt 110 fighters, while the faster, nimbler Messerschmitt 109s fly high to keep the British fighters at bay. The bombers are able to take their time over the target and carpet bomb the area.

Arthur Thompson, an electrician at Supermarine, is in his garden enjoying a rare day off when the sky fills with German aircraft. He watches with mounting horror as they turn toward their target:

> All I saw was round about a hundred bombers with a first class fighter escort hovering and hoovering above as cover and the 'aak-aak' or shell fire was doing no harm at all. Apparently they started over

Bournemouth, but I watched the aircraft come over the Forest, over the factory, drop their bombs and away over Portsmouth and away back home.[20]

Around seventy tonnes of bombs drop from the Heinkels. Seven bombs hit the Woolston works and one rips open the Itchen factory. Air-raid shelters are hit again and many more bombs fall on surrounding homes and the gasworks across the River Itchen.

Joan huddles in the basement of her office building as the harsh roar of aircraft engines is overwhelmed by the low thud of high explosives:

> We fell to the floor and once again suffered the onslaught of bombs while the building shuddered – I found myself repeating the words of a hymn – 'Faint not Fear, his arms are near' over and over again.[21]

Slowly, unsteadily, Joan stumbles from her temporary shelter into the acrid air outside:

> I was met with the sight of the factory in ruins under a pall of dust and smoke. The canteen was in flames.[22]

She rushes up Sea Road past the skeletons of homes to the nearest public shelter – the cellar of the Red Lion pub:

> This shelter was full of women and children singing shakily . . . 'Hear him pecking out his melody, peck, peck, pecking in the same old key' – The Woody Woodpecker Song. I shall think of this day every time I hear this song in future.[23]

When the all-clear siren finally sounds, Joan bumps into her father rushing from his shipyard toward the factory. He'd been told that Supermarine was in ruins.

Fifty-five people were killed that day and Spitfire production ground to a sudden halt.

As each British fighter is shot from the sky, there will soon be nothing to replace it. The bare metal bones of aircraft are just visible, crushed among the smoke and twisted metal of the factory. Workers gradually drift back to the ruins to see what can be salvaged. By the time metallurgist Harry Griffiths arrives, recovery teams are already there – searching not for bodies, but for surviving machinery and for those precious Spitfire parts:

> The whole of the roof had been smashed to pieces and they were putting the work that was salvageable into boxes and, while they were at it, they were throwing the debris out of the windows. And they were singing their heads off. I don't know if it was to keep their courage up but they were working like the devil.[24]

Just one Heinkel and two Messerschmitt 110s are shot down by British fighters. Two Spitfires and one Hurricane are lost in dogfights with the 109s far above the city.

Reports of the raid reach the Ministry for Aircraft Production in Whitehall before the bombers have turned for home. The news has been relayed from deep inside the factory, from supposedly bombproof telephone boxes. As each air-raid begins, volunteers sit inside the steel boxes, grimly nicknamed 'coffins', and slam the door shut behind them. It's not a job that young Cyril Russell envies:

At half a dozen points inside the factory were positioned conical steel structures. They stood about seven feet high, were about one yard diameter at the base and two feet diameter at the domed top cover. A door, hinging outwards, took up roughly a third of the circumference, and slits of about six inches long by half an inch wide were cut to eye-level at four positions. [25]

The volunteers survive the raid to telephone in their damage reports. Despite the smashed roof and a vicious carpet of shattered glass, things aren't quite as bad as they first appear. Three Spitfires nearing completion have been destroyed, another twenty or so damaged, but crucially, the main machine tools are unharmed.

The Minister for Aircraft Production, Lord Beaverbrook, takes the call in Whitehall, steps into his chauffeur-driven car and speeds straight down to Southampton. Beaverbrook knows Britain's air defences are at breaking point and he needs to survey the damage for himself.

Beaverbrook is a newspaper baron. A tightly wound ball of ego and energy, he's turned the failing *Daily Express* into the country's best-selling newspaper. In the process he's cultivated a powerful set of friends and confidants from among Conservative politicians. Valuing his reputation for getting things done at any cost, Prime Minister Winston Churchill has just appointed him as Minister of Aircraft Production. Beaverbrook has a clear but far from simple brief – to rapidly increase the supply of aircraft coming out of Britain's factories. The loss of the Woolston and Itchen works is an enormous blow. Supermarine's quiet, analytical General Manager, H. B. Pratt, has been badly injured in the bombing so Beaverbrook is met at the factory by a frustrated and angry young firebrand – twenty-eight-year-old Works Engineer Len Gooch:

> That evening Lord Beaverbrook arrived here at the works and I was sent for to see this wizen-faced little man and, all of a sudden he started thumping the table and I wasn't in the mood to be thumped very much on that occasion, having had two days of this, and he thumped the table and said, 'Look, I want more Spitfires, more Spitfires!'[26]

Len and his team are battered, bruised and tetchy, but they do have something to offer Beaverbrook. The factory is a clear and obvious target, so why not move the machinery out, to places the Germans can't spot? A plan has already been drawn up, to move vital machinery out of the factory and into three car-repair workshops nearby.

The idea catches Beaverbrook's imagination. But it doesn't go nearly far enough. He barks at Gooch, 'You're going to disperse all over the country. Complete dispersal.'[27]

It's an extraordinary idea that makes little sense to Gooch. In Germany, the great rival to the Spitfire, the Messerschmitt 109, is being built in just 4,000 man hours by specialist teams on a state-of-the-art production line – much faster than the bombed Southampton factories. Beaverbrook now has a very different 'handmade' vision.

Production of the Spitfire must restart immediately, but not in a city-centre factory. It will be 'dispersed' –

hidden among towns and villages. Hidden among the very people the Spitfire is supposed to protect.

It will be a change that breaks every rule of modern industry, and places thousands of new workers right in the heart of the battle. No one quite knows how it will work, but the promise is made. On the BBC Home Service, Beaverbrook makes a solemn speech to the nation:

> We must recognise that the enemy is making preparations for the invasion of Britain. Invasion by land and sea, but principally by the air. How are we going to meet and overcome this danger? By industry, by enterprise, by endurance and by fortitude. And here let me say that all those engaged in the production of aircraft have demonstrated a vast measure of industry and endurance. We have the inspiring knowledge that the aircraft we make will be directed by firm and strong hands. It is our task to assure the pilots that the supply is adequate to necessity and worthy of the men who fly for Britain over the land and over the sea.[28]

2

The Phoenix

An odd sense of relief shuddered through the workforce. The death and destruction was deeply traumatic, but the tension had been released. For Cyril Russell and his friends, the months of waiting for the inevitable, months of fearing the worst, were over:

> I went back into the works to collect my tools, and saw for myself that men were busily uprooting the vital jigs, while the almost undamaged Machine Shop was actually working, under a missing roof.[1]

What Cyril saw wasn't proof of life; it was the factory's last spasm before death. Spitfire production on the banks of the River Itchen was at an end. Production had to be restarted from scratch, out of the Luftwaffe's reach, and twenty-year-old Cyril Russell was set to play a crucial role:

> On the following Monday I was informed by the Labour Exchange to report to Hendy's Garage, just off Southampton High Street, and upon arrival, there were fuselage jigs that had been uprooted from Woolston, and reset in what had been, on the previous Thursday, one of Southampton's largest garage and car showrooms. So in five days after what had looked

like a disaster, I was back building Spitfire fuselages, and I was by no means the only one, or this the only working unit.[2]

Three garages had been scouted out before the bombing and some machinery and parts transferred in expectation of the destruction. Hendy's Garage in Pond Tree Lane and Seward's Garage in Winchester Road were big two-storey buildings with concrete floors connected by a ramp. They would be ideal for fuselage production. Lowther's Garage in Shirley would be a tool room for the jig borer machines. It was a start, but nowhere near enough to replace the two abandoned factories in the city's industrial heart.

The Spitfire is a complex, cutting-edge piece of technology, requiring thousands of parts made by highly skilled workers. The idea that production could be parcelled out across southern England, entrusted mostly to an untrained workforce, was, on the face of it, preposterous. The Spitfire's rival, the Messerschmitt 109, is a similar aircraft in many ways, but its construction is much more straightforward. Both push the boundaries of 1930s aircraft design – they're constructed from the new light but strong aluminium alloy, duralumin, both boast retractable undercarriages and both house their pilots in enclosed cockpits.

These planes were a huge step forward from the wood and fabric biplanes that formed much of the RAF's front-line force right up until the late 1930s. In engineering terms, they're an advance even on the Hawker Hurricane, the backbone of Fighter Command, with their innovative monocoque construction.

Essentially this means that the stresses and strains of the aircraft are supported by the all-metal external skin. This does away with the need for internal struts and wires to maintain the shape and integrity of the aircraft in flight. It's a construction technique that should make the aircraft easy to make on the kind of production line used for cars and large consumer goods. That's certainly the route Messerschmitt took at their giant Regensburg factory in Bavaria.

On the Messerschmitt 109, the fuselage frames are integrated with the outer skin and pressed as a large piece of metal. Around thirty pieces could then be bolted together to create the aircraft. The Spitfire, by contrast, required an almost pre-industrial approach, with hand-crafted elements forming around 300 major pieces that had to be fitted together in the final assembly process.[3]

Even in the purpose-built factories on the River Itchen, this was a slow process, significantly slower than the production lines of Bavaria. Recreating the complex processes in dozens of small workshops around southern England would prove to be an enormous logistical and technical challenge.

So how was it to be done? How could production be dispersed and hidden away from German bombers? Those decisions were being made in the Polygon Hotel. The upper floor of Southampton's top hotel, much loved by variety stars playing in the Empire Theatre next door, was taken over by Supermarine management and the men from the Ministry.

There was Wilf Elliott, the small and rotund Works Superintendent, Wing Commander Kellett from the Air

Ministry, and the spiky and ambitious Works Engineer, Len Gooch. They were joined by two of Lord Beaverbrook's own men, sent from the Ministry of Aircraft Production to keep an eye on proceedings, a mismatched pair called Whitehead and Cowley.

Those were the officials making the decisions, but it was Denis Webb, the charming but impatient Spare Parts Manager, who had to kick-start the dispersal. Brought up by the sea in Hartlepool, Webb was obsessed with boats and planes from a young age. When he refused to join the family law firm, his father reluctantly supported his application for an apprenticeship at Supermarine, then best known for building flying boats. With the company since 1926, he'd seen it grow from a rich man's hobby to a vital cog in the British war effort.

Webb's first job after the bombing was apparently simple – find more premises in Southampton that can receive the surviving jigs and machine tools and get some kind of Spitfire production up and running as soon as possible. Things started pretty smoothly:

> I was sent to see the Manager of the Sunlight Laundry in Winchester Road, which was on the outskirts of Southampton, to tell him we were about to requisition his Works and ask him to move out as quickly as possible. I don't remember any great opposition, although rather naturally the idea wasn't popular, but they soon began to move all their machinery out.[4]

George Goldsmith, owner of the laundry, put up no resistance. The laundry machinery was moved out within a few days, but Spitfire production couldn't restart there instantly:

Sunlight Laundry had got all their stuff out by the weekend, but we found that the roof trusses were absolutely deep in cotton lint, which was festooned everywhere, giving me the answer as to why nothing lasted long when sent to a laundry. It took us all quite a while to get it all cleared off![5]

Winchester Road, five miles from the bombed factories on the main route to London, proved a happy hunting ground for Webb. Just next door to the Sunlight Laundry was the Hants and Dorset Bus garage. It possessed the holy grail for Webb – a high ceiling. High enough to house a double-decker bus, and high enough for the enormous frames – or jigs – used to construct the Spitfire's distinctive leaf-shaped wings. The only problem was, someone had got there first:

> My next task was to go and see the Deputy Town Clerk, whose name was Bernard Fishwick, and ask him to shift all the Trailer Pumps of the Fire Brigade out of the Hants and Dorset Bus Garage.[6]

Trailer pumps are small fire engines that can be towed behind a truck. The whole of the bus depot was filled with these pumps, surrounded by heavy sandbags.

> Fishwick refused to clear them out, on the basis that the Fire Brigade Trailer Pumps were of more importance to the town than 'bloody Spitfires'. My argument that 'bloody Spitfires' in adequate numbers could make trailer pumps unnecessary was not accepted, and so I said the matter would have to be referred to Beaverbrook and Company, who would undoubtedly enforce their removal![7]

Lord Beaverbrook's representatives in Hampshire, Cowley and Whitehead, proved themselves willing and able to use their emergency powers to overrule local officials and enforce the seizure of premises:

> The Deputy Town Clerk had removed his trailer pumps by the weekend, if I remember correctly, but all the sand bags were still there and we were trying to find out from him where we could dump them – it being anatomically impossible to put it where we would have liked.[8]

Webb's travels around Southampton next took him to the back of a cemetery:

> Someone told me that there were a row of empty huts behind Hollybrook Cemetery, which might be useful for a Finished Part Store headquarters, and I had a look at them and agreed.

There was only one problem. The huts weren't connected. To move equipment from one to the other, you'd have to go outside, letting light flood out into the dark of the blacked out city. Webb wasted no time in coming up with a solution:

> I went to a local builder in Winchester Road and got him to build a brick passageway, to link up all the Huts at Hollybrook Stores, so that at night there could be free movement between the buildings without having to have light traps on the doors. While in the builders' office drawing up plans, a woman clerk started to complain that, with Supermarine

in Winchester Road, they would get bombed now!
I said, 'Well Woolston had most of it up to now and
surely in wartime we should share and share alike.'[9]

As Webb left the office, he muttered some rude words
about this lady, who turned out to be the builder's wife.
This was a tense time in Southampton. It was technically
still a town rather than a city and the business community
formed a tight-knit web of personal, financial and political
connections. When weight was thrown or wives insulted,
the word would spread quickly. Denis, though, was a man
on a mission; he didn't feel any particular need to be liked.

By the time Denis had finished his work, the Winchester Road area was a hive of activity. Alongside
Supermarine's new workshops in the Sunlight Laundry,
Hants and Dorset Bus Depot and Seward's Garage, were
numerous small, independent companies. Allom Brothers
Lighting was producing electrical components for the
Spitfire. Round the corner in Emsworth Road, Auto
Metalcraft was turning out fuel tanks and air ducts. Even
the smallest businesses in the area converted to Spitfire
production. Tucked behind a derelict mansion at the junc-
tion with St James's Road was a car repair garage called
Light and Law. Don Smith, a fourteen-year-old apprentice
at Auto Metalcraft, was sent round there with a message.
Hearing an electrical hum, he followed the buzz upstairs.
There sat a solitary old man crouched behind a lathe,
'with a great big heap of parts on the floor beside him'.
He seemed to ignore the interruption, so Don nervously
asked him what he was doing:

Well, excuse me if I carry on working, son, I'm
making parts for the Spitfire propeller and I'm the

only one left doing this. If I don't continue then we aren't going to win the Battle of Britain.[10]

While Denis Webb drove the streets of suburban Southampton, bullying and cajoling rightful owners out of their premises, his bosses' eyes were trained on a map pinned to a bedroom wall in the Polygon Hotel. It showed roads, spreading inland like the fingers of a hand, from Southampton out to a sixty mile radius. There simply weren't enough suitable buildings in Southampton to produce the Spitfire and the city, with its dockyards and surviving factories, was still a prime target for German bombs. The Luftwaffe may not be able to target Seward's Garage or the Sunlight Laundry in the same way they could pinpoint the Woolston and Itchen plants, but raids were increasing in intensity and workers were frightened for their lives.

Lacking the deep shelter of London's underground stations, the people of Southampton had very limited protection from German bombs. Housing was tightly packed around factories and docks, making them vulnerable to high explosives and the incendiary bombs that would inevitably follow. The medieval wine vaults beneath the Old Town were pressed into service for city-centre residents, while those lucky enough to have gardens could build their own corrugated-iron Anderson shelters. Others could huddle in the cellars of pubs or simply hide under their beds and hope for the best.

What Southampton does have in abundance, however, then and now, is wild countryside on its doorstep. City residents could finish their day's work, pick up some food for their dinner, and trek into the Hampshire Downs or the New Forest, from where, on the worst nights of the Southampton Blitz, they watched their city burn.

To some outsiders this smacked of cowardice. Germany's radio propagandist, Lord Haw-Haw (born William Brooke Joyce), was heard referring to the 'trekkers' in his broadcasts to Britain, portraying them as symbols of a civilian population cracking under pressure. To the Nazi leadership, it seemed only a matter of time before public opinion would force Winston Churchill to negotiate a surrender. To the trekkers themselves, their nightly walks were a perfectly logical response to relentless raids. In the summer, autumn and early winter of 1940, Southampton suffered fifty-seven raids in total, with one fierce firestorm even visible across the English Channel in Cherbourg. The shock of the September raids on Supermarine persuaded the family of shipyard worker Maisie Nightingale to join the nightly trek:

> We were going to Bursledon at night and coming back here to Southampton in the mornings. Any vehicle, anything you could get hold of to pick us up, to bring us in. And that's what we were doing morning and night, going right out of Southampton. And I can remember Lord Haw-Haw saying, 'we know what you're all doing, trekking to Bursledon out of the way'.[11]

Initial briefing documents prepared for the War Cabinet in Whitehall maintained that morale remained high in the bombed industrial cities, but a visit to his Southampton parishioners by the Bishop of Winchester revealed a different picture: 'I went from parish to parish and everywhere there was fear.'[12]

*

Lord Haw-Haw, it seems, had hit upon a painful if unsurprising truth – bombed civilians are frightened. It was a fact confirmed by despatches from the reporters of Mass Observation. This social research project was set up by three Cambridge University graduates in 1937, with the intention of recruiting a panel of volunteer observers to systematically record human activity around Britain. The documentary filmmaker Humphrey Jennings, the South African communist poet and journalist Charles Madge, and the anthropologist Tom Harrisson had observed the abdication crisis of Edward VIII and Mrs Simpson with great interest. They were interested in how the views of the man or woman in the street might differ from the representation of public opinion in the popular newspapers of the day. Their radical response was to actually ask people what they thought. Mass Observation asked volunteers to keep diaries and answer questions about the moral and political issues of the day. How did they feel about the triumph of Fascism, for example, or the truth of horoscopes?

It began as a poorly-funded amateur research project, but when war broke out Mass Observation took on a new importance. For once the government was genuinely interested in what the public thought. How would they respond to rationing or conscription? Were propaganda films believed or dismissed? How long might the British people be willing to fight on alone?

The government set up the Home Intelligence unit of the Ministry of Information in February 1940, led by the pioneering BBC television producer Mary Adams. Her team was tasked with writing weekly reports that tracked the shifts and turns in public mood. They issued their own questionnaires and gathered evidence from Regional Intelligence Officers, and even through postal intercepts, but it was the work of Mass Observation that gave them their most acute insights into public opinion. The unit and the higher echelons of government could use evidence from Mass Observation's diaries and specially commissioned reports to spot weaknesses in vital public services and monitor the state of national morale.

Mass Observation sent its own young south-coast reporter, Leonard England, to gather public opinion in the wake of the raids that followed the Supermarine bombing. On 4 December he reported 'a general feeling that Southampton was done for' and observed, at first hand, a partial evacuation of the devastated city:

> Throughout Monday there was apparently a large unofficial evacuation. Two people spontaneously compared the lines of people leaving the town with bedding and prams full of goods to the pictures they had seen of refugees in Holland and Poland. Some official evacuation took place on the Monday, but at the Avenue Hall rest centre a group of fifty waited all the afternoon for a bus to take them out.

From 4.30 p.m., England watched a stream of people leaving the town for the night. As well as capturing his own observations, England drew on the accounts of other Mass Observers:

When Mr Andrews left the train at the docks, he was impressed by the seeming deadness of the town; there were no cars, and hardly any people except those that had left the train with him. But farther out people were moving. The buses were full, men and women were walking with their baggage. Some were going to relations in outlying parts, some to shelters, preceded by their wives who had reserved them places, and some to sleep in the open. 'Anything so as not to spend another night in there.'[13]

There were hitch-hikers on the roads out of town and trains full of women and children. Some seemed determined to leave for good but others had little baggage with them, as if they would return soon:

In some neighbourhoods whole streets had evacuated, most people leaving a note on their doors giving their new address; one such notice read 'Home all day, away all night'.[14]

Deeply concerned by these transcripts, the government despatched the Inspector-General of Air-Raid Precautions, Wing Commander John Hodsoll, to report on the situation in Southampton. He returned to London with an ugly portrait of a local authority out of its depth, led by a mayor and town clerk he seems to have regarded as incompetent. Petty local rivalries hampered rescue efforts, while the failure of fire-fighting equipment cost time, property and lives. At the height of the intense night raids in November, hundreds of fire-fighters were drafted in from surrounding districts, but incompatible equipment and a lack of water made their job almost impossible.[15]

When these twin reports later came to the attention of the local *Southern Daily Echo* newspaper in the 1970s, the response was explosive. Letters to the *Echo* honed in on Leonard England's apparent ignorance of the town and his failure to travel much beyond the town centre. His Mass Observation report was slammed as 'a slander on the town'. That 'slander' was magnified just a few weeks later when the Public Records Office released the Hodsoll report to public scrutiny for the first time. Furious local worthies set up the Defence of Southampton Committee to robustly refute the accusations made in the report. Feelings ran so high that the Home Office, under pressure from a local MP, agreed that a statement of rebuttal from the people of Southampton would be filed alongside the Hodsoll report in the National Archives.[16]

Fifty years on from the 1970s, it seems astonishing how much anger was provoked by these reports. Southampton suffered repeated devastating attacks from the Luftwaffe. Photographs reveal the utter ruin of much of the city centre and docks. An observer described 'a blazing furnace in which any living thing seemed doomed to perish'. To hear that some citizens were eager to leave, to read the opinion that many could talk of nothing else and seemed 'dangerously near neurosis', seems entirely unsurprising. It speaks volumes for the power and persistence of the myth of the 'Blitz spirit' that these observations provoked such anger. Alongside bravery there was fear. Alongside stoicism there was confusion, panic and even incompetence. Perhaps today it's acceptable to admit that.

What is truly remarkable is that each morning, even at the height of the Southampton Blitz, so many workers *did* return. Despite the failure to protect the city's citizens, despite the disruption to transport, telephones, gas, electricity and water, despite having to bury their dead and find new homes, those who could walk picked up their tools each morning and did their best to keep the Spitfires, ships and machine tools coming.

To Lord Beaverbrook, however, it had become obvious that Southampton was simply too vulnerable to be relied upon. Spitfire production had to be dispersed far from the dangers of the city.

3

From Hairdressers to Riveters

Cyril Russell, the smart young panel beater, was back building Spitfires within days of the factory bombing. Salvaged fuselages were dragged to Hendy's Garage nearby, a back-up base that had already been prepared for production. Cyril worked on those until the rest of his team was gathered together at Seward's Motors on the edge of the city.

Things were pretty chaotic there, with dumped, unsorted components rescued from the wrecked factory strewn across the floor. The equipment Cyril worked with was relatively light and mobile, so his team became one of the first to abandon the city completely. Six miles north of Southampton is the hamlet of Chandler's Ford. On the edge of the village was an agricultural showroom filled with tractors, ploughs, sacks of fertiliser and crates of wellington boots. The owner had no intention of moving out, so the Requisition Officer brought in a labour gang under police escort to remove all the farming machinery piece by piece. On 22 October, just one month after the factory bombing, Cyril started work in the freezing cold but relative safety of a cavernous warehouse in the Hampshire countryside.

The return of the German bombers over Southampton in the November Blitz emphasised, if any reminder were needed, that the dispersal had to be accelerated. A series of enormous raids between 22 November and 1 December devastated the city. Only one of the dispersed production sites was shut down by the bombing, but the Blitz had a significant impact on production. The supply chain from independent engineering companies was disrupted and workers had to take time off to deal with damaged homes or injured relatives. The Vickers-Armstrong Quarterly Report neatly summarised the impact:

> During the heavy air attacks on Southampton on the nights of November 30th and December 1st, none of the dispersal premises suffered damage, but they were affected for two weeks by the partial failure of electricity and gas supplies and great inconvenience was caused by complete breakdown of telephone communications. Employees experienced serious hardships due to bomb damage to their houses and difficulties of transport also affected the number of hours which could be worked.[1]

With a renewed sense of urgency, Denis Webb and other senior Supermarine staff began to radiate out from the city on the hunt for suitable buildings.

They were urged to take note of dimensions and type of construction, as well as services such as electricity and water, and to report back as quickly as possible. As the information came in it was collated and recorded by Works Engineer, Len Gooch, who made the decisions on which buildings to seize.

The owners were telephoned and the intention to requisition made known. If any obstruction occurred, the 'big guns' were brought to bear, in the shape of Cowley or Whitehead, known to the rest of the team as 'Beaverbrook's Bully Boys'. Denis Webb worked closely with them, rather relishing the mixture of fear and fury they provoked in the stubborn owners of workshops and showrooms:

> Cowley was a great character – Canadian – always calm and unruffled and with a great sense of humour. I think we all liked him immensely. Whitehead was a different kettle of fish and I don't think anyone cared for him much.[2]

Businesses weren't always keen to give up their precious space to the Ministry of Aircraft Production, but Beaverbrook's Boys were there to twist arms, and twist they did. One warehouse in Winchester had already been requisitioned by the Ministry of Food, but Cowley had other plans for it. Denis couldn't resist listening in to Cowley's end of the telephone conversation:

> 'What have yer got stored there?' went the slow Canadian drawl. Silence, then 'Pineapples?' with a rising inflection of exasperation. 'Jesus! Do you think you can win this goddamn war with PINEAPPLES?!'

The result was inevitable:

> We got the premises the next day. Cowley would cheerfully have requisitioned any building we wanted, or even one we didn't want!

The irrepressible Cowley once gazed out of the window of the Polygon Hotel to eye up the elegant classical curves and columns of the city's pride and joy – its brand new venue for music and dancing, the Southampton Guildhall: 'Gee, that would make a swell erecting shop . . .'

Cowley was keen until the local men managed to convince him that a polished sprung dancefloor would not be suitable for Spitfire manufacture. And when the Mayor of Salisbury objected to the requisitioning of the Wiltshire and Dorset bus garage, Cowley was straight on the phone:

'Tell me, Mr Mayor,' said Cowley, 'I believe you are the Patron of the local Spitfire Fund, is that true?'

'Oh yes,' came the obviously very proud reply

'Well,' said Cowley, with his inimitable drawl, 'May I suggest that you close that Fund and start another, to erect a Statue to the Mayor who thought Spitfires weren't necessary!'[3]

Capitulation, of course, followed soon afterwards.

By mid-November 1940, a total of around thirty-five separate workshops, large and small, were up and running and the number was rising rapidly.

As the dispersal spread and distances grew, five distinct 'production centres' emerged from the initial chaos, with the largest one, Southampton, at the base of the fan. Twenty-five miles to the north-west there was a production hub in the cathedral city of Salisbury, with fuselages made in Wessex Motors, wings in the Wiltshire and Dorset Bus Company, and sub-assemblies of various kinds in Anna Valley Garage.

A purpose-built factory was thrown up, forty miles to the north-east, as part of the Newbury hub. In nearby Reading, Vincent's Garage, opposite the busy railway station, turned out fuselages, while the Caversham factory across the River Thames installed the Rolls-Royce engines.

Fifty miles north-west of the city, Trowbridge became the final production centre and it was there, to Fore Street

Garage, that Cyril Russell was transferred again, to lead a team of Southampton veterans and new local workers. He'd been promoted several times since the war had started – not bad for a lad who'd left school at fourteen.

Gradually, Cyril, and many more like him, created order from chaos.

Each production centre ran more or less autonomously, with full production capability; there were workshops and supply stores, and each one also had an airfield, sometimes tiny with rough grass runways, where the planes could be assembled and then flown for the first time.

By 1942 there would be almost 250 sub-contractors, supplying over fifty workshops and stores, all across the English counties of Hampshire, Wiltshire and Berkshire. The upheaval for the owners of requisitioned buildings was enormous. But it wasn't just the premises that had to adapt. The workers had to find their place in this new system too.

For the survivors of the Itchen and Woolston factories, that meant finding transport to their new place of work or a nearby place to stay. Cyril Russell's experience was typical. For the first few weeks of his time at Chandler's Ford, he was able to get a lift to and from his Southampton home with a colleague lucky enough to have a private car and sufficient petrol coupons to keep it on the road. As winter drew on and night bombing raids became more frequent, however, the commute began to feel more like a dance with death. Driving through pitch-black countryside after a twelve-hour shift toward a city sky criss-crossed by searchlights, just to spend the night in an Anderson shelter, quickly lost its appeal.

As a single man, lodging could usually be found. Billeting Officers would scout out spare rooms and force homeowners to rent them to essential workers. Families, though, were more of a problem. Many Supermarine workers didn't want to leave their loved ones in a bombed city while they slept in the relative safety of Salisbury or Trowbridge. Finding family-sized homes was tough, and some at Hursley Park, where the design departments were dispersed, found themselves living in conditions that resembled one of the grimmer novels of Thomas Hardy.

The Bell family, for example, were bombed out of their modern three-bed house in Southampton and forced into a single room in Glebe Cottage, an isolated farm building at Farley Chamberlayne, miles from the nearest village. Supermarine engineer Bill Fear was billeted with his wife and children in a tiny timber cottage shared with the Wild family and their saddlery business. Much of the area around Hursley had no running water or electricity until the 1960s. When you wanted water, you had to trek to a hand-pump and that pump had to be primed – a slow and painful job first thing on a frosty morning. The primitive toilets still needed a visit from the 'Night Soil' man to be emptied.

As late as March 1941, the Vickers-Armstrong's Quarterly Report for Supermarine described: 'great difficulty encountered in transferring skilled men from Southampton because of the impossibility of getting sufficient and suitable housing.'[4]

The problem of accommodation was one of the biggest factors preventing the successful completion of the dispersal. One solution, employed at several different dispersal areas, was to erect prefabricated housing close to the workshops. For the staff at Hursley, a group of around eighty temporary prefabricated homes, known as hutments, was constructed at Hiltingbury, on the outskirts of Chandler's Ford.

For the Bell family, these hutments were a lifesaver. Eileen, then a child of four, recalled their time at Glebe Cottage and the move to the hutments:

> There were four in our family. Mum, Dad, myself and my brother Leslie and we all lived in one room. But we weren't there that long. It was months more than years, 'til as soon as the hutments were put up. We were the first ones, actually the first ones to move into the hutments. On Hook Road as you went down, it was the last one before you came to the bungalows. My father wanted the end one because we had quite a large garden there compared to anybody else. Apart from the fact we were desperate because we were homeless.[5]

Isolated from most services and public transport, the hutments were only intended to last for ten years, but persistent local housing shortages meant that they continued in use until the late 1950s. A close community developed, but, as Eileen recalled, conditions remained pretty spartan:

> They were cold and it was damp too. They were sort of two layers of asbestos with straw in the middle. On the inside walls there was a sort of a cardboardy stuff, hardboard. I say hardboard, probably didn't have

hardboard in those days, so a sort of thick cardboard. It was very damp and very cold, but I guess we just got on with it, I suppose.[6]

For some of the younger Southampton workers without families, the dispersal gave them opportunities they could never have dreamt of before the war. Ken Miles was a quick, clever and very cheeky fifteen-year-old when war was declared. His shopkeeper father told him that it was time for him to leave school and get a job. Ken decided that he wanted to work in engineering, so an uncle with trade union connections at Supermarine secured him a position as an Apprentice Aircraft Fitter at the assembly works at Eastleigh airport, on the edge of the city.

On his first day at work, another lad proudly informed him that he himself was an Apprentice Aircraft Engineer, who would go on to work in the Design Team, while Ken could only hope to be a lowly Fitter. Ken wasn't having any of that: 'Thinking things over, I decided that I did not want to be an aircraft fitter but an aircraft engineer, so banged on the door of Mr Nelson, the Manager.'[7]

Ken was instantly hussled out of the office for his cheek. A few days later a black limousine pulled into Eastleigh. Out stepped the elegantly pressed figure of Chief Test Pilot Jeffrey Quill, accompanied by some senior RAF officers and the Supermarine General Manager, H. B. Pratt. They were there to examine progress on a new Spitfire variant being assembled by a select group of workers. Ken hopped on to the wing, tapped Mr Pratt on the shoulder and asked to talk to him about his employment prospects. Ken was lifted off the wing and whisked into Mr Nelson's office for a final warning.

Ken, though, wouldn't let his ambitions lie and the company was eventually forced to concede, sending him to their Work School, where he gained the extra qualifications needed to train as an engineer.

The unpredictability of war then changed his life once more. His parents left Southampton to escape the bombing, so Ken returned home to act as a fire-watcher for his father's shuttered electrical business. A German bomb narrowly missed the shop, but the windows were smashed in and Ken was forced to move in with relatives in nearby Romsey.

His workplace at Eastleigh airport was no safer. Late on the afternoon of 11 September 1940, a small group of aircraft approached the site at a low altitude. They were taken to be RAF Blenheims arriving for repair. By the time they were identified as Messerschmitt 110 fighter bombers, it was too late. They dropped their bombs on the Cunliffe-Owen aircraft factory next door to Supermarine. Ken made it to the nearest shelter just in time, but his best friend was among the fifty-two dead. A few days later the Woolston and Itchen factories were bombed. Ken was given the toolbox of a neighbour killed in that raid.

With the dispersal now in full swing, Ken moved from place to place, wherever the need for a relatively experienced Southampton hand was greatest. At Hursley Park he worked on experimental fuselages; at Seward's Garage back in Southampton, he took over from a draughtsman who was joining the RAF and found himself drawing jigs and tools. From there he went to Supermarine's new Trowbridge base to learn the Robinson Process. This involved creating templates of parts on aluminium sheets.

They had to be accurate to within a 5000th of an inch. Ken was pretty pleased with his first attempt, but it was rejected by the inspector, as were his second and third attempts.

A senior draughtsman eventually took pity on Ken and told him that his standard wooden ruler could never achieve the necessary level of accuracy. He needed a stainless-steel ruler graduated to a 100th of an inch. Ken called in at Lawson's, his local hardware shop, to ask for a Chesterman No.761/3 rule. The shopkeeper told him that they'd sold out at the start of the war and been supplied with no more since. Ken, in his typically forthright fashion, composed a polite letter to the Minister of Aircraft Production about his plight: 'I received his very nice reply, which had me to return to Lawson's, where a rule would be waiting.'[8]

Ken got his new ruler, and met his future wife at Trowbridge, but he fell out with the manager and was transferred to one of the Supermarine sub-contractors, Shorts in Winchester, where he started his own Drawing Office. By 1944 he was back in Southampton, working on milling machines at Lowther's before another move to Chandler's Ford. There he was appointed as Assistant Liaison Officer, travelling around the dispersed factories and workshops, solving the numerous problems that such a complex operation threw up. By then he certainly knew his way around this extraordinary business.

The Spitfire odyssey of Ken Miles shows just how difficult it was to staff these dispersed factories. Experienced Southampton survivors were the gold standard, but they couldn't make enough Spitfires on their own. Some were injured in the bombings, some had joined the armed forces, others had found better paid work close to home.

The rest were spread thinly across the three counties and regularly moved to wherever they were needed most.

To support their efforts, a new generation of aero industry workers had to be found and trained, both in the city and in the depths of the English countryside.

Most able-bodied men aged from eighteen to forty-one who hadn't been conscripted into the armed forces were already engaged in essential war work, so women were the principal target of this new recruitment drive. Pat Pearce moved from Salisbury's Marks and Spencer to the new Supermarine base at the Wilts and Dorset Bus Company. Ordered into war work, she was given the choice of the Women's Land Army, growing the food for a hungry besieged nation, or the grease and banter of Spitfire production. The factory option sounded warmer and drier, but it was the extraordinary noise of the production line that hit her first, as she told the makers of the documentary film *Secret Spitfires*:

> I shall never forget the day I started. It was frightening, really. There were all these people, ready to start on their jobs, and we were right over the other side and we all had different little jobs. But then it got so noisy you could hardly hear yourself speak![9]

The long dayshift hours – 8 a.m. to 7 p.m. – certainly hurt and there was absolutely no slacking on the line. Nightshifts were a little more forgiving. With no foreman watching, there was the occasional chance to grab a cigarette break or even a lie down.

Bette Blackwell moved from hairdressing to riveting. Rivets were a defining feature of the Spitfire. In the quest for incremental improvements in speed and manoeuvrability, the design team had fixed the prototype aircraft together with flush-headed rivets rather than the cheaper mushroom-headed rivets, which were easier to fit in a production line system. Supermarine's aerodynamicist, Beverley Shenstone, had seen the German Heinkel 70 fast mail plane at the pre-war air show in Paris and run his hand along its fuselage. It was so smooth that he thought it might be made of wood. Shenstone wrote to the designer, Ernest Heinkel, to ask how he'd done it. He received an utterly charming reply, which explained the use of coun-tersunk rivets and several layers of paint.

Under pressure to compromise on cost, the design team conducted flight tests in which split peas were stuck on top of the flush rivets to simulate the drag factor of a conventional rivet. The split peas slowed the aircraft down by a full 12 mph. Flush rivets would therefore have to be fitted to the most important front-facing panels, making life that bit tougher for Bette Blackwell and her friends.

Riveters could work singly, using a toggle gun, closing up the rivet from both sides of a metal plate, or in pairs, using the reaction riveting system. One worker would be inside the aircraft body holding up a block while their partner fired in the rivets from outside the fuselage. Bette was one of the smaller girls, so usually found herself screwed up in a little ball inside the Spitfire.[10]

'Rosie the Riveter' is one of the iconic images of World War Two and a powerful feminist symbol, portraying the new industrial workforce of women in the US, the UK and the Soviet Union as strong and vital components of the war effort. Sure enough, despite the noise and the vibration, riveting was regarded as one of the plum factory jobs for the Spitfire women. One of the places where the riveting compressors roared away night and day was the requisitioned Sunlight Laundry, on Winchester Road in the Southampton suburbs. Two women keen to work there were sisters Florrie and Daisy.

Florrie and Daisy Snelling had been working for British American Tobacco, filling packets by hand. With practice and nimble fingers, they could pick up a bunch of cigarettes and know they had twenty. They heard there was work to be had at Sunlight Laundry, making Spitfire parts. The work was more interesting, the pay was better and, as it turned out, there was ample opportunity for romance.

Florrie met Roy, a sheet metalworker, and Daisy fell for Reg, a toolmaker. Courtship was quick and engagement soon followed for both couples, giving Florrie and Daisy's father a very expensive headache. He couldn't afford two weddings, so the girls saved up their coupons for their wedding dresses and organised a joint ceremony.

In wartime, opportunities for joy and celebration were scarce, so, when they came around, everyone joined in. Reg and Daisy and Roy and Florrie got married in June 1942 at the local church and they were amazed when they emerged into the light of a summer Saturday afternoon to find that the road was full of their work-mates from the Sunlight Laundry and the other dispersal sites along Winchester Road. They'd all been given the afternoon off, to see the Spitfire Sweethearts come out of St James's.

The newlyweds couldn't have their reception at the parish hall, just across the churchyard, as it too had been requisitioned, and was being used to make Spitfire fuel tanks, so they found somewhere else for their double wedding reception, with their double bridesmaids, their twin cakes, and a shared first dance.

A newspaper sent a photographer and the next day's article was headlined 'Spitfire Sweethearts', a name that stuck with them throughout their lives. Reg and Daisy still have a memorial stone at St James's, outside the church where they were married.

After the wedding, it was back to work. According to Daisy's daughter, her mother was given a special piece of equipment made by Reg that helped her to rivet much faster than the other women in the factory. Supermarine took a dim view of the homemade tool and put a stop to Reg and Daisy's daring innovations, but for a while at least she was the ace riveter of the Sunlight Laundry.[11]

Southampton workers like Florrie and Daisy were familiar with the latest industrial processes, but in some rural areas, Supermarine's arrival brought a first taste of modern technology. Horizons were widened and lives changed.

Essie Dean was delighted the day she finished her education at tiny Keble school in Hursley. She'd never really enjoyed her lessons and now, at fourteen years old, she could pursue her dream job – milking cows on a local farm. Her step-father, Alfred, had other ideas, though. A job at Supermarine would bring the family a lot more money than a milkmaid's wages.

A junior clerical position was arranged and poor Essie, who was only truly happy in the quiet fields around her home, was issued with a military pass and told to clock-in with the forbidding commissionaire at Hursley House's grand entrance hall. The new base for the design team had been the stately home of the local lords of the manor.

Had she been born a year or two earlier, Essie might well have found work there as a chambermaid or among the kitchen staff. Now the oak-lined corridors were busy with draughtsmen and engineers developing new marques of the Spitfire.

Essie's first job was in the ballroom, entering information into a ledger. For the painfully shy Essie, the hustle and bustle of this enormous room was agony. One of the older men would scream and suffer a full-blown panic attack if anything was dropped on the floor. As a survivor of the factory bombing, he was almost certainly suffering some form of post-traumatic stress disorder, but for a four-teen-year-old country girl it was terrifying. Essie would return home every night crying to her mother. Alfred offered to pay for shorthand lessons, but fortunately a kindly secretary taught Essie to type and she was able to quickly move on to more interesting work.

The gilt drawing room, with its tall Palladian pilasters, was another majestic space, but the cheeky cherubs peeking out from swags of foliage and the grand imperial eagle gazing down made Essie smile. This really was the hub of the whole dispersal process. Every part of the Spitfire had a code, a number that uniquely identified the mark and sub-assembly all the way down from complete aircraft to individual rivets. For the dispersed works to make their Spitfires, the right code sheets had to be sent to the right workshop. This was Essie's job, one she turned out to be very good at. It was an all-male section, but she impressed them with her quiet efficiency and, as she sped documents up and down the grand staircase from the Drawing Office to the Planners, she got to know just how the Spitfire was made.

Once her typing skills improved, she moved next door to the light and airy conservatory, typing the specification sheet masters of designs of individual aircraft. She was paid nineteen shillings a week, ten of which she passed straight to her mother for housekeeping. That left her plenty of spare cash for camp dances and for the bus fare to the cinema in Winchester. When thousands of American troops arrived in Essie's neighbourhood in preparation for D-Day, it was obvious that there was no going back. The quiet, country life with its strict hierarchies and moral boundaries was over for Essie and so many young men and women like her. Like it or not, their world had opened up.[12]

*

As the dispersal gathered pace, the workforce continued to grow, and grow. All sorts of odd jobs needed doing, and help came from both ends of the social spectrum.

Park Place is a country house, very close to the little town of Wickham, about fifteen miles east of Southampton. It was the home of Sir James Bird, who had briefly owned Supermarine before it was taken over by the Vickers engineering corporation. Sir James came out of his comfortable retirement to offer his management expertise – and his land – to the Spitfire production team in their moment of need.

Bird had an extension built to his home – a very long, narrow hut. It was where women, mostly from the Women's Voluntary Service in Wickham, came every day to sort rivets.

Rivets, like so many small Spitfire parts, came from many different suppliers. There were different colours, different lengths, different shapes. Flat rivets for the leading edge, domed rivets to the rear, heat resistant rivets for under the mighty exhaust. Millions of tiny rivets. And they all need sorting by these volunteers, all woman over fifty, many much older. Construction of the Spitfire had truly become a common cause.

In this shared endeavour of Spitfire production, one group of women felt a particularly heavy weight on their shoulders. These aircraft were incredibly complex pieces of machinery and those built in dispersed factories were the product of many hands in many different workshops across southern England. It was the job of the Aeronautical Inspection Directorate to make sure that the finished aircraft were safe to fly. The RAF's most precious resource – the lives of its fighter pilots – was in the hands of women like Dorothy Handel.

Dorothy had travelled to England from her home in Melbourne, Australia, in 1927 and, at the age of fourteen, entered domestic service. She worked as a kitchen maid and cook, passing through several stately homes before being dismissed from South Lodge in Horsham for daring to step out with a young man. By 1939 she was an Aircraft Woman 2nd Class, working as a cook for the Women's Auxiliary Air Force at Hendon. A chance meeting with a manager from the Aeronautical Inspection Directorate alerted her to an opportunity for some much more satisfying war work.

After a month of basic training, she joined Folland Aircraft, inspecting small parts and components at their factory at Hamble in Hampshire. From there she moved to the Cunliffe-Owen factory in Eastleigh on the edge of Southampton, where Spitfires were assembled for delivery and damaged aircraft repaired. It was here that Dorothy, busy inspecting the internal construction of a Spitfire wing, caught the eye of a photographer from the *Portsmouth Evening News*. With her platinum blonde hair scraped back, waistcoat and fitted pin-striped trousers, the image perfectly captured the calm professionalism of the A. I. D. women. This rare photograph from inside a Spitfire factory illustrated an article that dubbed Dorothy and her colleagues 'Heroines of the Home Front':

> Their busy fingers toiled with a feeling for their work which only an artist can feel. And as I saw them tripping around the factory with their torches and test equipment, testing the wings, fuselages and installation of Halifaxes or Spitfires or sitting in the cockpits of Hudsons testing the controls, I, at least, felt sure that whether or not the Battle of Waterloo was won on the playing fields of Eton, the coming Battle of Europe at any rate will be won above all in the factories where these girls, ever trim and ever smiling, work night and day.[13]

The smitten journalist's tone may have been somewhat patronising, but Dorothy's bosses were full of praise for their new workforce: 'It is not an easy or a light job, but the way these girls have taken it up is simply marvellous . . . You know I will not change one of my girls for a fully qualified man.'

By late 1943, some 500 women were working for the Aeronautical Inspection Directorate in the region around Southampton and Portsmouth. Many left secretarial jobs or domestic service to take on these highly technical roles.

The dispersal of Spitfire production had a social and economic impact Lord Beaverbrook could never have imagined. Women across the region gained technical skills and scientific expertise at a speed that would have been impossible in peacetime. Career paths that only a very determined few had followed before were now open to all.[14]

Supermarine didn't just bring opportunity to rural England, it also brought danger. When production first began to move away from Southampton city centre, there were genuine worries that the towns and villages being converted into Spitfire factories would now become targets for the Luftwaffe. Cyril Russell, survivor of the devastating September raids, certainly felt that the war was following him wherever Supermarine sent him:

> We had a few scares from the German raiders when bombs were dropped in the general area, by accident or by design, and one night, a night fighter caught a bomber less than a mile east of us, and it exploded in the air. It was such an explosive force that we wondered how any night fighter could escape the blast.[15]

Despite the immense effort expended in relocating the Spitfire works out of the German line of fire, Britain was engaged in total war. Nowhere was truly safe. And sometimes the Luftwaffe punched through Britain's defences, hitting the new factories by accident or design.

One of the smaller dispersal sites was in the sleepy medieval market town of Bishop's Waltham, eleven miles from Southampton. It was an unlikely target for a German attack – a Victorian Brickworks.

Supermarine had commandeered buildings at Blanchard's Brickworks. They moved in, taking over drying sheds and one of the kilns. When German bombs hit the brickworks, Denis Webb and General Manager H. B. Pratt were sent to investigate:

> We drove up to the entrance to the works where the usual group of gawkers had gathered, plus some police, and air raid wardens. I was somewhat dismayed to see a notice: Keep Out, Unexploded Bomb! Pratt turned to me and said, 'All right with you, Webb?' I croaked out, 'I suppose so', and we went in.
>
> All the parts were undamaged and we could see that almost immediately, but Pratt went on poking around wanting to see where the bombs had gone in, while I did my best to stop wetting myself! I was heartily relieved when we left the site.[16]

Many people around the town witnessed the bombing, and there are a few contemporary accounts, one from a young lad who was herding some cows, another from a lady named Betty. They all mention that when the bomber came in very low, they could hear it, and see from its markings that it was a German plane, and when he dropped his bombs, the sky turned red. The reason for that explosion of colour was the brick-dust that had accumulated over the years, flying up into the air, turning the Bishop's Waltham sky crimson.

*

The dispersal of aircraft production from two purpose-built factories on the banks of the River Itchen to dozens of separate buildings – from laundries and glove factories to stately homes – changed the lives of many. Hairdressers became riveters, milkmaids now knew their camshafts from their carburettors, and modern life – and the war – arrived with a sudden jolt in some of the sleepiest corners of southern England. But did it work? Did Beaverbrook's vision of a 'handmade' aircraft built by the people to defend the people prove its worth?

On an individual basis, there's no doubt that ordinary people – the trained workers from Southampton and the new rural workforce – gave everything they had to build their Spitfires.

By 1942, Cyril Russell had moved to a tiny hangar at High Post Flying Club, where all the various parts from across three counties and beyond were being assembled into a plane:

> The hangar was just large enough to take three Spitfires, providing the first two had their tail ends pushed into the opposite corners, so the third could get between them. Its wings then blocked the other two in. The push-open sliding doors were just high enough to permit the propellers to clear at the top, as long as the three blades were in the Y position. For canteen facilities we had to trek half a mile across an airfield.

It was utterly unsuitable for Spitfire construction. But, as with all the workers building the secret Spitfires, Cyril and his team made it work:

> It used to take roughly two and a half days, of day and night shifts, to make a single Spitfire, and with three on the go together, the overall effect used to be six aircraft per week produced.[17]

Hardly enough to turn back the Luftwaffe, but Cyril's six Spitfires a week were joined by around 750 from Reading and Newbury, 2,300 from Salisbury, 600 from Trowbridge and well over 4,000 from the factories of Southampton.

Production from the dispersed factories would eventually be overhauled by the Castle Bromwich production line, which turned out a total of around 12,000 Spitfires by the end of the war. In sheer numbers, those were vital aircraft, but Castle Bromwich lacked flexibility. If a new variant was needed to combat a fresh German threat or to take on a specialist role, such as aerial reconnaissance, the dispersed workshops were able to adapt much quicker than the modern production lines of Birmingham – or Bavaria.

The Big Plan had been rolled out with remarkable speed and efficiency. The great dispersal of Spitfire production, in the autumn of 1940 and beyond, had transformed lives and sent hundreds of desperately needed fighters into the air. Hidden away in small, scattered communities, thousands of people took on the risk of becoming a target themselves to build the plane that would help turn the tide of the war. It's certainly an heroic tale. But one big question remains – was dispersal the correct strategy?

Denis Webb, Supermarine's Manager of Spare Parts, was well placed to answer that question. Ever vigilant, ever curious, ever critical of stupidity and waste, he was scrutinising the official reports, taking notes and coming to his own conclusions: 'During the first quarter of the Dispersal, output fell by over 50%, averaging under 14 Spitfires a week.'

The company's Quarterly Report for March 1941, six months after the raid, reported: 'considerable retardation of production.'

If we were to stop history here, then it would have all have been for nothing, but Denis Webb goes on to tell a different story. It's a story of rivets and screws, a story of wings and propellers and a Merlin engine – built, assembled and ready to fly. It is a story of thousands of tiny parts, each one essential to make a complete fighting aircraft.

And it's a story of thousands of people, each with their part to play. And thanks to these people, eventually, it came together:

> We had delivered our 1,198[th] Spitfire by the time we were bombed out. By early 1941, the first stage dispersal was complete and all requisitioned laundries, garages, bus stations, glove factories, steam roller works and strawberry basket works were producing Spitfire parts and components. This first stage took six months. By the end of 1941, we had got back to pre-dispersal output, and by the middle of 1942, we had surpassed it.[18]

Looking back years later on his role in this extraordinary story, Denis was determined that credit for its success should not go to management pen-pushers or government ministers but to the ordinary people across Hampshire, Wiltshire and Berkshire, who willingly put themselves on the front line:

> The legend which seems to have built up about the splendid forward planning which made the dispersal

such a success really needs correcting. I can say, quite definitely, that it was NOT pre-planning that made the dispersal successful, but the quite marvellous improvisation and hard work that did the trick.[19]

The workers, old and new, had turned disaster into triumph, but this logistically complex dispersal was ruinously expensive in terms of cash as well as human lives. Two factories had been abandoned and dozens of new ones built or re-fitted. The government needed cash to pay for the dispersal and for new Spitfires to replace those being shot down in the Battle of Britain. Fortunately, making money was the specialist subject of one particular government minister.

4

Paying for the Spitfire

The heroes and heroines of this book were mostly in their teens or early twenties when they made their impact on the story of the Spitfire, but even those too young to fly, fight or flourish a rivet gun could do their bit to build Britain's greatest fighter.

There's twelve-year-old John Masters, for example, who tells the BBC Home Service that he's already raised £4 10s to build Spitfires:

> I'd lived for seven years out in East Africa, in fact my father's out in Palestine now. I've got quite a lot of interesting things that we've brought back from there, so I held an exhibition of them in a tent in our garden, and charged people a penny or tuppence to come in. Then a man who sold poultry promised me a dozen eggs and I raffled them at sixpence a time.[1]

There's nine-year-old Pamela Weeks from Liverpool, who asks the children of Britain to support her fundraising efforts. She especially wants other girls named Pamela to contribute to a Spitfire:

> My daddy is a soldier, and if we can win the war quickly all our daddies will soon be with us. At night

when we have to go to the shelter it is no use being grumpy . . . none of us mind if we know there are lots of planes to chase Mr Hitler. Even a penny will help.[2]

Eight-year-old Patricia Boncy has a particularly cunning plan. She offered 15 shillings to the Minister of Aircraft Production with an explanatory note:

When my mummy has taken me out and I have wanted to use a public convenience she has had to pay a penny. So I thought if we did the same at home it would help your fund.[3]

Patricia, Pamela and John were all early recruits to a national fundraising army. They were pioneers of an innovative form of crowdfunding that would prove to be one of the war's most successful propaganda campaigns. It is the Spitfire Fund that turned a camouflaged tube of aluminium, steel and Perspex into an icon of resistance, a precious tool in uniting the nation and persuading the public at home and abroad that Britain really did stand a chance against the might of Nazi Germany. Those who wonder why the Supermarine Spitfire became the hero of countless films, war comics and books while the Hawker Hurricane is largely forgotten will find the answer in the story of the Spitfire Fund.

*

On 10 May 1940, German forces invaded Belgium, Luxembourg and the Netherlands. The Phoney War was over – this long anticipated *Blitzkrieg* was clearly the precursor to a full-scale invasion of France. Britain was in grave peril, but her Conservative government was deeply split. Calls for a coalition 'national' government to face the challenge grew louder. The opposition Labour Party refused to support the incumbent Prime Minister, Neville Chamberlain, but they would agree to serve under an alternative Conservative choice. The only plausible candidates were Foreign Secretary Lord Halifax and First Lord of the Admiralty Winston Churchill. Halifax accepted that Churchill was in a stronger position to unite the nation and by that evening the King had appointed Winston Churchill as the United Kingdom's new leader.

Before the week was out, Churchill acted to accelerate the production of the aircraft that he knew would be vital to the defence of Britain. A new government department was created, the Ministry of Aircraft Production, and a new minister was appointed to lead it – Max Aitken, Lord Beaverbrook.

A more controversial appointment to such a sensitive position would have been hard to imagine. The Canadian press baron had bought the *Daily Express* and turned it from a minor player to a bestseller, using it shamelessly to promote his own political interests. He was hated by the Left and distrusted by many on the Right. Beaverbrook had been an ardent opponent of war against Germany and supported Neville Chamberlain's policy of appeasing Hitler, right up until the Nazi invasion of Norway. Despite this he was a close personal friend of the arch-enemy of appeasement, Winston Churchill.

Beaverbrook's grandson, Maxwell Aitken, explained their surprising bond to the BBC:

Perhaps they shared some traits, being regarded often as unorthodox and un-establishment. I think Churchill saw a lot of himself in my grandfather. There would be huge rows between them occasionally and they wouldn't speak for a month or two. And then they would be back as best of friends.[4]

On taking office, the new Minister wasted no time. Production had to be increased at every British aircraft factory and fighters were the urgent priority. Planes, though, were expensive bits of kit and every arm of the military was pressing the Treasury for more funds. Beaverbrook needed an idea that would push his Ministry to the front of the queue. The answer lay with the British public, but the path to their wallets took a circuitous route around the Empire.

The initial idea was inspired by an innocent enquiry from a reporter on *The Gleaner*, then as now, the bestselling newspaper in Jamaica. Observing the loss of men and machinery at Dunkirk, the journalist asked the Ministry for Aircraft Production, 'How much is a bomber?' A Ministry official, presumably feeling that he had more urgent business on his plate than providing accurate answers to colonial journalists, offered up a random figure of £20,000. Soon another enquiry came through to the Ministry from the Caribbean. This time it came from Sir Harry Oakes, an American goldminer who had moved to the British Bahamas to evade tax. He asked the Ministry, 'How much for a Spitfire?' The Ministry again plucked a figure from the air – £5,000.

[Oakes, incidentally, was brutally murdered three years later. He was killed in his island home with a miner's hand pick, his body burned with insecticide and then covered with feathers from a mattress. The investigation into this grisly crime was led by the Governor of the Bahamas, the Duke of Windsor and former King Edward VIII. Unsurprisingly, given the Duke's reputation for arrogance and incompetence, the murder inquiry was a shambles. Oakes's son-in-law was put on trial for murder but acquitted when it became apparent that two Miami detectives hired by the Duke had fabricated evidence. Despite numerous reassessments of the evidence and recreations of the case by writers and film-makers the murder of Sir Harry Oakes remains unsolved.]

In 1940, Oakes was a newly ennobled baronet with influential friends in the Caribbean, Canada, the US and the UK. When his cheque arrived, it caught the attention of Ministry officials and Lord Beaverbrook himself. Could the public be persuaded to act like Sir Harry and pay a little more than their taxes for the war effort?

A notional price list was drawn up and the public invited to contribute. A completed Spitfire, ready to fly, was priced at £5,000, but it wasn't just whole Spitfires – everything was given a price. The Merlin engine alone was £2,000, a spark plug was 8 shillings, and a petrol tank was £40.[5]

As the Battle of Britain raged in the skies above southern England, the idea caught the public imagination. They may not be able to fly the planes themselves, but they could do their bit. As Lord Beaverbrook's grandson pointed out to the BBC in *Spitfire: The People's Plane*, the Fund performed a dual role:

Quite clearly the whole country was faced with a dire financial situation. But it wasn't just financial, it was morale, it was getting the population to connect with the efforts that were being made in the name of war – connect the nation with the production of aircraft. And it certainly made them feel like they were contributing.[6]

With one rivet priced at sixpence and sixty screws for five shillings, a Spitfire was within the reach of most pockets and donors knew that whatever they sent could make a difference. Young Peter Bottomley knocked on doors and shook his collecting can around the streets of Guildford. When he sent in his contribution, he got a thank-you letter by return from Buckingham Palace. It informed Peter that he'd raised enough money to buy a tyre for the rear wheel of a Spitfire. Every time he heard a Spitfire in the Surrey sky, he could look up and wonder if that was his tyre he could see.

Up and down the country, in towns and villages, community groups, families and workplaces, people threw their money into jars and buckets with the aim of putting one more Spitfire into the air. Beaverbrook appointed a public-relations officer to co-ordinate the scheme and send out the thank-you letters. The BBC Home Service began broadcasting news of major donations and local and national newspapers revelled in the fundraising stunts and heart-rending stories:

> A concert, which two little girls from Emscote got up last Friday brought in 4s for the fund. The 'artistes' sang and danced and also provided lemonade. An excellent example![7]
>
> A German pilot gave a five-mark note on Sunday to the Mayor of Chatham's Spitfire Fund. The pilot

who had been shot down by RAF fighters over Kent earlier in the day was being escorted under armed guard by train through Chatham.[8]

The Spitfire bracelets presented by Messrs P. H. Woodward and Co have realised £8 17s 6d for the Fund.[9]

The Misses Warner, of Boston Spa, have given money in memory of their brother, Lieutenant John Weston Warner D. F. C, who was killed in action in the last war.[10]

To keep the public engaged, individual Spitfires were named after the communities or businesses that supported them. There was a Wakefield Spitfire, a Walsall Spitfire and a Warner Brothers Pictures Spitfire. The Kennel Club named their adopted plane Dogfighter.

One measure of the Fund's hold on the national imagination was the viral spread of 'The Spitfire Song'. It was composed in 1940 in the Spitfire's home town of Southampton by Horace Maybray King, an English teacher at Taunton's School. It was originally performed as the 'Hampshire Spitfire Song', but Spitfire Funds across the country latched on to its popularity and rewrote it accordingly, creating the Northampton, the Tyneside or the Yorkshire 'Spitfire Song'. A year later the country's most popular big band, Joe Loss and his Orchestra, cut a disc and turned it into a chart-topping hit, with singer Sam Browne belting out the patriotic message with gusto:

> There's no need to take cover
> When you hear these engines sound,
> British planes are in the skyways,
> On their daily vigil bound.

We'll make fun of their number,
Write our name upon the wing;
When the planes are flying over
You will hear all Britain sing!

The perky tune was arguably more inspiring than the lyrics, but the 'Spitfire Song' certainly added to the Fund's popularity from one end of the country to the other. The song's composer, perhaps wisely, didn't pursue a musical career. Horace Maybray King was elected as a Labour MP for Southampton and, in 1965, went on to become Speaker of the House of Commons.

*

At the tiny museum in Market Lavington in Wiltshire, they still have the records of their village's very own Spitfire Fund.

Little Susan Dark gave all her pocket money of sixpence to the fund, and guests at the local hotel had to put a penny in the jar every time a German plane was shot down. There was a dance competition that raised £1, 5s and 6d. At the Green Dragon pub they had a picture on a table, an outline of a Spitfire measuring a metre across. Drinkers were asked to fill it in, with valuable silver coins in the Spitfire itself and copper coins to make up the landscape in the background. Once filled in, that came to about £8.

In total, Market Lavington raised nearly £90 and sent it off to Lord Beaverbrook. It was nowhere near the £5,000 it would have taken to buy a Spitfire, but larger towns in the county contributed more and a local Spitfire eventually took to the air, proudly named the Wiltshire Moonraker.

The pilots of these aircraft sometimes got an unexpected boost from the towns and societies that funded them. Group Captain Wilfred Duncan Smith, father of the future Conservative Party leader Iain Duncan Smith, flew a Spitfire Vb called 'Crispin of Leicester'. He reported receiving 'woollen socks, mufflers, fruit cakes and love letters' from the women of Leicester. One package included a long scarf in the colours of Leicester City Football Club and a proposal of marriage.[11]

Meanwhile, in the north-east of Scotland, the people of the fishing town of Arbroath set about their fundraising with particular gusto. At a public meeting on 30 August 1940, it was unanimously agreed to establish the Arbroath Spitfire Fund. Even before the fund had officially opened, three of the burgh's teenage girls had done their bit. Sheila O'Brien, Ishbel Herron and Ella Webster, all of Airlie Crescent, called at the Burgh Police Office and handed over a small bag containing 35 shillings. They reported that the money had been raised from concerts they had held at the back doors near their homes, and hoped that the money might help towards victory in the war. The entertaining trio expressed their fervent wish that the money go towards a Spitfire.

It wasn't just local stars who sang and danced their hearts out. The biggest music hall act of the day, Sir Harry Lauder, was persuaded to play a charity concert, and a male voice choir of refugees from the Polish Army treated the town to some rousing folk songs. There was a boxing tournament, a police dance and a generous cheque from exiled Arbroathians in the United States.

By March 1942, £4,869 19s 11d had been raised. The Town Council agreed to make up the difference and two months later a brand new Spitfire Mark Vb EP-121 was delivered to RAF Burtonwood in Cheshire.[12] They called her the Red Lichtie, after the beacon lit by the monks of Arbroath Abbey, which once shone across the harbour, guiding the fishermen home. The log book from the Red Lichtie survives, so the people of Arbroath can still trace the adventures their pennies paid for.

From Cheshire, Red Lichtie was allocated, in May 1942, to 501 Squadron. With them she flew 'Rhubarb' and 'Roadsted' missions. 'Roadsted' was the Royal Air Force parlance for attacks on coastal shipping, while 'Rhubarb' missions were ground attack operations into Occupied Europe. Small formations of aircraft, usually just a pair, flew at low-level into France or Belgium to seek out and shoot up opportunistic targets on the ground. Rhubarbs were normally flown in bad weather so as to avoid enemy fighters, but this made them extremely dangerous, with numerous navigational and mechanical failures as well as losses to anti-aircraft fire.

The Red Lichtie and her pilots somehow survived those dangers and she was transferred to 131 Squadron, where she flew 'Circus' missions. These were raids in which small groups of bombers attacked railway junctions, ports or factories. The ground targets were largely irrelevant, as the bombers were there as bait to lure shy German fighters into the sky, where they could be jumped by large wings of Spitfires. By now the Germans had effective ground radar and became adept at picking and choosing which formations to jump with their own fighters. Any British pilot shot down was almost certain to be captured, putting them out of action for the duration of the war.

These 'Rhubarb' and 'Circus' missions were designed to pin Luftwaffe squadrons down in Western Europe, helping ease the intense pressure on the Soviet Union in its life or death struggle with Nazi invasion. It's debatable whether that particular goal was achieved, but viewed purely from Fighter Command's perspective, overall losses far outweighed the damage caused to the enemy.

The Red Lichtie gave its all in these raids on Occupied Europe. After one particularly lively sortie, Sergeant Strang limped home, landed on the wrong runway and collapsed the Lichtie's port wheel. Her radiator shield was ripped to shreds on a later high-speed dive while Squadron Leader Johnnie Johnson, flying her on 13 February 1943, shot down a Focke Wulf 190 south-west of Boulogne.

Her final posting was with 412 Squadron of the Royal Canadian Air Force. On 29 June 1943, piloted by Sergeant W. H. Palmer, she was approaching her home airfield in Lincolnshire when she stalled. Palmer struggled to restart the engine and she spun earthwards, smashing into the ground. Astonishingly, Sergeant Palmer survived the crash, but his Spitfire did not.

The Red Lichtie was beyond repair, but the people of Arbroath wouldn't let their plane be forgotten. A replica aircraft was commissioned, which now takes pride of place at the entrance to the Air Station Heritage Centre in the neighbouring town of Montrose.

*

The story of the Spitfire Fund is largely one of pocket money, pub games and charity football matches, but the growing glamour of the Spitfire's reputation pulled cash in from all across the globe.

Campaigns in Britain's colonies raised enough to have whole squadrons named after them. There was No.74 (Trinidad), No.167 (Gold Coast) and No.114 (Hong Kong) squadron. From India, the Nizam of Hyderabad sent enough cash for two Spitfire squadrons.[13]

Even the Belgian Congo got in on the act, issuing postage stamps featuring a Spitfire superimposed over a jolly native scene. Money raised in the notoriously harsh colony helped buy a batch of Spitfires, including one flown with 350 (Belgian) Squadron by a Belgian RAF pilot, Henri Picard. The plane was named Luvungi after a Congolese town. Native people at the time were being forcibly conscripted into the mining and rubber industries, enduring foul working conditions on behalf of the Allied cause. Their white Belgian overseers enjoyed a profitable war, selling rubber to the British while clandestinely smuggling diamonds to the Germans. This presumably provided enough francs to buy those postage stamps for the Spitfire Fund.[14]

Much more fun was the rum-fuelled fund-raising of the Fellowship of the Bellows. In October of 1940, a group of expatriate Brits met in a hotel room in Buenos Aires and decided to do their bit for the mother country. The RAF Museum in Hendon has a collection of their colourful membership cards and booklets of rules. Museum curator Andrew Dennis says that the Bellows was a sort of jocular secret society – 'Funds through Fun' was their motto. The odd name came from the idea that bellows create a 'force of air' or, if you prefer, an Air Force. As with all private members' clubs, there was a strict hierarchy. From your first donation, you were designated as a 'Wiff'. A complicated formula based on the number of German aircraft shot down and the member's original pledge gave you the chance to progress through the ranks. A 'Wiff' could become a 'Puff', then a 'Gust', with the ultimate aim of reaching the heady heights of 'Hurricane' status. Any encounters with fellow members in normal life entailed a secret hand gesture and the correct greeting, 'Hello Fellow Bellow'.

From the evidence of their distinctive sense of humour, it should come as no surprise that the founders of the Fellowship of the Bellows were mostly rather wealthy men with careers in banking and advertising, living enviable, sun-soaked ex-pat lives in South America. The Bellows were a long way from the ration books and blitzed buildings of Britain, but they were determined to play their part and the image of the Spitfire perfectly fitted their racy lifestyles. These were not people shaking cans in the avenida, they were organising charity golf matches, canasta sessions and sophisticated balls and cocktail parties where the cash flowed as liberally as the Negronis.

Local interest boomed and the ex-pats were soon outnumbered by Argentinian supporters, pushing the membership up to around 60,000. From there the Bellows spread their net among the smaller British migrant communities of Paraguay, Uruguay, Brazil and Mexico. The Brazilian 'Bellowship', as they liked to call themselves, bought eight Spitfires and later a squadron of Typhoons. The accounts of the Fellowship held at the RAF Museum reveal that, by 1945, the generous socialites had raised nearly 10 million Argentinian pesos for the cause, the equivalent of £600,000.

From the UK and abroad, £13m was raised for the war effort – equivalent to £650m today. There were hundreds of named Spitfires protecting the skies above Britain and fighting in every theatre of war.[15]

Lord Beaverbrook could see just how effective the Spitfire Fund was proving for the Treasury, for national morale and for his own standing in government. He took to the BBC to give thanks:

> We have had a flow of contributions come in, all of them sent to us for the purpose of buying aircraft. Some of the gifts are large, and some of them come from people of limited means.
>
> We value the cheque for £25,000, but we value, too, the gift from the Telephone Operators of Winchester, who sent us 38 shillings, to buy screws for a Spitfire. I can tell those generous men and women that the screws will be well and truly praised.
>
> And I want to give my warmest thanks now to all those, who by their contributions have given inspiration and encouragement to the Aircraft Ministry. And here let me say, on behalf of the aircraft industry, that we will try to face with fortitude the ordeal of battle.[16]

So, did the Spitfire Fund keep the Nazis from the door and our pilots in the sky? Absolutely not. In terms of the overall cost of the war, £13 million was a tiny drop in an enormous ocean. During World War Two it's estimated that half of Britain's annual Gross Domestic Product of around £300bn was spent on defence.[17] The money raised from the Spitfire Fund went straight to the Treasury, not into the Spitfire factories. It may have been spent on ships, rifles or wages. But the money was never the real point. The fundamental purpose of the Spitfire Fund was to rally the nation, to unite the people in the fight against

Hitler. The toughest months of German bombardment provoked plenty of resentment – class struggle, looting and profiteering. It would have been no surprise if those at the sharp end of the Home Front – in the flames of London's East End or Clydebank's shattered docks, say – had begun to question why this war was being fought and why it was only them that appeared to be expendable.

The Spitfire was a stunning representation of Britain united. It provided living, flying proof that British workmanship could prevail against the Nazis, proof that the daily sacrifice being asked of Britons would one day prove worthwhile. The Spitfire was a bulwark against defeatism and the Spitfire Fund helped enrol the whole nation in that fight. What else could unite cocktail-sipping sophisticates in Buenos Aires with Scottish fishermen and Home Counties schoolgirls? What else but the Spitfire?

5

The Other Spitfires

The story of the dispersal and the secret Spitfire factories emerged long after the war was over, but it dovetailed neatly with the Spitfire legend that had already become a pillar of British identity. The Spitfire Fund and the plane's starring role in films, books and war comics had placed it front and centre of Britain's wartime saga. Hearing the stories of the people who built it under the toughest conditions, in the quaintest of places, simply added to the romance of the Spitfire. Our war was won, it seems, by a beautiful machine hand-built in a rural idyll. What could be more British than that?

Supermarine's dispersal of production is certainly an astonishing tale of improvisation, heroism and quick thinking under fire. But there's another Spitfire construction story that displays other, less attractive elements of the British character – class conflict, arrogance disguising ignorance, and an inability to accept responsibility for failure.

Not all of the RAF's Spitfires came from the shattered Southampton plant or the workshops spread across the towns and villages of southern England. A huge purpose-built factory, perched unromantically on the edge of Birmingham, built more than half of the Spitfires ever flown. In the story of Castle Bromwich there are few heroes but plenty of plausible villains.

In 1976 a BBC TV documentary crew caught up with a group of workers from the Castle Bromwich Spitfire factory. It betrays a little about the prejudices of the BBC at the time that they are the only people in an hour-long film whose names are never given. This cheerful group, obviously pleased to see each other after thirty-five years, told presenter Raymond Baxter about the long hours and the tough work they had all endured. But every one of these men and women, most in their teens and early twenties in 1940, was immensely proud of the Spitfire and their role in its construction:

> When you went through C block, which was the final shop, and to see them lined up ready to ferry across the road for testing, it was a wonderful sight.
>
> It wasn't just putting nuts and bolts together, you felt a lot of satisfaction.
>
> You knew that things had got to be perfect or as near perfect as possible, and you made a special effort.[1]

It's hard to reconcile those first-hand recollections with a report from June 1940 written for the Ministry of Aircraft Production by Sir Richard Fairey:

> In my opinion the greatest obstacle to an immediate increase in output is the fact that labour is in a very bad state. Discipline is lacking. Men are leaving before time and coming in late, taking evenings off when they think fit . . . In parts of the factory I noticed that men idling did not even bestir themselves at the approach

of the Works Manager and the Director who were accompanying me.[2]

A factory designed long before the outbreak of war to 'shadow' the work of the vulnerable Southampton factory had produced not a single Spitfire by the summer of 1940. When new fighters were most desperately needed, Castle Bromwich had nothing to offer. The roots of the crisis lay in decisions made years before these young workers took their place on the new production line.

*

By 1935, Hitler had abandoned any pretence of abiding by the Treaty of Versailles ban on German re-armament. The strength of the Luftwaffe was no longer a secret and the British government felt compelled to respond. The country's safety would depend on the latest fighters and bombers, but Britain's aeronautics industry was still rooted in an era of craftsmanship rather than production lines – the country simply didn't have the factories to build them fast enough. The government's solution to this problem was the Shadow Factory Scheme.

The Shadow Scheme aimed to get mechanics, engineers and managers from Britain's thriving automotive industry to 'shadow' aircraft factories, so they could learn how to make aeroplanes. The government would build and equip new factories, and the motor car companies would staff them. The word 'shadow' simply meant that the two industries were to work in lockstep, not that this was a secretive, shadowy scheme.

Shadow Factories, run by car companies, sprang up all over the country. Rolls-Royce established one in Glasgow,

Rover set up in Solihull and Ford in Manchester. All made aeroplanes, or parts for aeroplanes. And for the best fighter of all, the biggest factory of all – a brand new state-of-the-art Spitfire production line was to be built in the West Midlands, in a village just to the east of Birmingham, called Castle Bromwich.

Supermarine in Southampton was expected to share its knowledge and supply some of its staff to kick-start production in Birmingham. Cyril Russell, the twenty-year-old Supermarine apprentice in the Southampton factory, watched the process unfold with more than a little envy:

> It was agreed that the Nuffield organisation would 'shadow' our Spitfire production. New, purpose-built factories were erected with all speed, and in Castle Bromwich's instance they had tooling — rubber presses, brake presses and the like — which the Woolston Supermarine factory had never known![3]

The Secretary of State for Air, Sir Kingsley Wood, handed control of the new factory to William Richard Morris, along with an enormous order for 1,000 Spitfires. Morris was a wealthy motoring magnate better known to the public as Lord Nuffield.

By 1938, then, the situation looked promising. A brand new £2.5m factory – the largest purpose-built aircraft factory in Britain – with government money behind it, and a champion of Britain's world-leading motor industry at the helm. But nothing happened. For two long years not a Spitfire was produced. The reasons are, even now, much disputed, but they certainly start with the man in charge.

Lord Nuffield was the founder of Morris Motors. Starting with a cycle repair business, he devised his own motorcycle before entering the cut-throat world of car manufacturing. A voyage to Detroit, home of America's motor industry, taught him the importance of the production line. Armed with good contacts and more than a little luck, he emerged as one of the winners in the post-First World War motoring boom. By the late 1930s, Morris was in his sixties, a rather eccentric figure who combined legendary frugality with odd, sometimes misplaced, moments of generosity. When Oswald Mosley began his swift and cynical transformation from Labour government cabinet member to Fascist demagogue, it was Nuffield who stumped up the cash for his new political party. Mosley, ground down by the tedium of a long dinner with Nuffield, sparked to life when a cheque for £50,000 was waved in front of him. Nuffield, though, wasn't handing it over without an irritating homily: 'Don't think my boy that money grows on gooseberry bushes. The first ten thousand took me a lot of getting.'[4]

Nuffield's undoubted expertise in mass production won him the opportunity to build the Spitfire, but his distaste for compromise and teamwork handicapped the project right from the start.

Construction of the factory was managed by the well-respected vice chairman of Morris Motors, Oliver Boden. Despite exceptionally wet weather in late 1938, the foundations on the greenfield site were laid quickly. By November, the first of the six big production buildings was going up, with work often continuing through the night under floodlights.[5]

Even before construction was complete, the eager Castle Bromwich team pushed ahead, using their new buildings to assemble aircraft parts sent from the busy

Woolston works in Southampton. War was imminent and the pressure to complete the factory was intense. On 20 June 1939, authorisation arrived for the supply of components and the Castle Bromwich Aircraft Factory was officially ready to go. All it needed now was a trained workforce and some cheerful co-operation from the people who already knew how to build a Spitfire – the hard-pressed team at Supermarine in Southampton.

A central figure in that relationship was Denis Webb, Supermarine's Spare Parts Manager:

> There was plenty of direct communication between Castle Bromwich and ourselves, via a teleprinter. It chattered away all day and half the night, asking for bits and pieces to overcome shortages, or asking for advice on drawing queries and such like, and did its best to drive me and others mad! There was plenty of direct communication and some of it was pretty rude.

Having spent millions of pounds developing the Spitfire, not everyone at Supermarine was eager to hand over commercially sensitive material to a rival company.

> If I had known what a short time the firm had been in operation, I and others might have been a little more patient, but our tempers and general nervous condition were a bit taut, caused by being near the coast and in the middle of a prime target, with frequent air raid warnings day and night. We were having a big enough struggle to build up our output of aircraft without worrying about Castle Bromwich.[6]

And then there was the workforce. Castle Bromwich had been chosen as a site for the factory as it's perfectly positioned in the Birmingham-Coventry-Oxford triangle that, then and now, contains the bulk of Britain's car manufacturing capacity, as well as the industry's component suppliers, with their expertise in tyres and oils, brake pads and metalwork. The workers around there knew all there was to know about building some of the world's best cars. What few of them knew anything about was making aeroplanes.

Engineering and assembly expertise was certainly available, but workers transferring from the motor industry needed time to get to grips with different scales, from the large wing assemblies to the minutiae of cockpit instrumentation. When production eventually got up and running, the manpower demands of the Armed Forces meant that skilled labour soon ran short and many of the replacement staff knew little or nothing about modern manufacturing.

Stan Woodley was moved from Supermarine HQ to help Castle Bromwich iron out its problems. He was shocked to meet the new recruits: 'The employees here were a complete mixed bag. We had lion tamers. We had variety artists. We had footballers. We had the lot.'[7]

Rumours about the difficulties at Castle Bromwich begin to filter back to Southampton. The author Leo McKinstry recounts a story from Supermarine staffer Bill Cox, who was chatting to an elderly fitter on the Birmingham production line: 'Make things with aluminium?' the fitter spluttered, 'Not bloody likely. That stuff is OK for pots and pans but we're going to make things to beat the Nazis. We'll use iron.'[8]

Apocryphal or not, the story shows how the Southampton team viewed their stroppier counterparts in the

Midlands. Young Cyril Russell worked himself into a fury about the reports he was hearing:

> Despite a massive order for 1,000 Spitfires, and the almost 'open cheque' to purchase materials, word soon passed back from our chaps there that the project was 'bugged' with industrial action (or inaction). The cumulative result ensured no Spitfires reached the flight testing stage.[9]

Founded by Noel Pemberton Billing, an eccentric independent MP with strong anti-Communist opinions and a very odd obsession with lesbians in public life, Supermarine had never been a company where workers' rights were high on the agenda. The powerful trades unions of the West Midlands quickly negotiated better terms of employment than their counterparts on the south coast.

Whoever or whatever was at fault – management, workers or government policy – the unfortunate fact was that Spitfires simply weren't coming off the production line. By the time Winston Churchill became British Prime Minister in 1940, there had been no Spitfires – not one – from the £2.5m factory. The farce at Castle Bromwich was fast becoming a national embarrassment. Churchill needed someone to shake things up. On taking up his newly created role as Minister for Aircraft Production, Lord Beaverbrook was made well aware that Castle Bromwich had to be turned around quickly. It wasn't just a matter of the planes themselves, although those were desperately needed, it was a question of local and national morale. His opening gambit was a Beaverbrook speciality, the poison-tipped telephone call. In his autobiography, *Out on a Wing*, the managing director of Morris Motors,

Miles Thomas, recalled the Nuffield end of this particular conversation:

> From the Cowley end of the line it quickly became apparent that Lord Nuffield was vociferously defending his Castle Bromwich organisation and making it abundantly clear that in his opinion the Minister of Aircraft Production could either have Spitfires or modifications but he could not have both. The moment of truth had arrived. Sarcastically, certain that he was putting the ace of trumps on the table, Nuffield shouted, 'Maybe you would like me to give up control of the Spitfire factory?' In a flash came the reply: 'Nuffield, that's very generous of you. I accept!' There was a click in the earpiece – the line went dead. Inwardly I breathed a sigh of relief. Nufffield's face was ashen. For a long time he did not say anything.[10]

The first part of Beaverbrook's hastily assembled plan had been a resounding success, but Beaverbrook would need some help with the next step. He made an urgent call to Sir Charles Craven, the Managing Director of Vickers, the parent company of Supermarine. There's no transcript of the call, but Southampton Spare Parts Manager Denis Webb was eager to believe the version he'd been told:

> Beaverbrook rang up Sir Charles, whose secretary said he was unavailable as he was attending to the wants of nature. This was no good for Beaverbrook, who sent a message back that Sir Charles must come to the phone immediately. On receipt of this second message, Sir Charles instructed his secretary to inform

Lord Beaverbrook that 'Sir Charles can only attend to one shit at a time!'[11]

Beaverbrook was presumably well used to reactions like that. He handed control of Castle Bromwich to Vickers in its entirety. Top talent from the original Southampton factory was poached to get things moving – Stan Woodley was one of the first to make the flit:

> Beaverbrook decided that he needed to have somebody with aircraft experience to produce some aeroplanes he required very, very quickly. I was sitting in my office in Southampton one day and I had a phone call, to say 'Pack a bag we're going to Birmingham'. And we walked in and took over the factory.[12]

The new Managing Director, Alexander Dunbar, wasted no time in implementing Southampton-style policies of hiring and firing. He wrote back to the owners of Supermarine at Vickers HQ:

> We are cutting our time in quarter payments for the staff, and I have spent a lot of time today arguing with the charge hands. Yesterday it was the Draughtsman's Union and last night it was the Progress Clerk, but it's all in a day's work! Incidentally, we are sacking at least 60 Jig and Tool Draftsmen next week. We have tried to find out what they were doing but the answer is 'Not A Lemon'. In the meantime we manage to build the odd Spitfire.[13]

At first slowly, and then at blistering speed, the fortunes of the Castle Bromwich factory began to turn around. As the group of young wartime workers recalled for the BBC in 1976, their workload mounted fast:

> We used to work, in those days, six days a week, and in the offices was 8 'til 6.
>
> In fact we all did those sorts of hours.
>
> And there were people who worked two or three nights on the trot with no break at all.
>
> Of course it was a heck of a night shift when you changed over from days to nights. You did 7 on Saturday morning 'til 12, and then you came on at 7 at night and you did 28 nights straight up. It was a month about in those days.[14]

Beaverbrook enjoyed his moment of triumph in the summer of 1940 when, a mere six weeks after dramatically sacking Lord Nuffield, ten completed Spitfires – the first ten, known as the 'Ten in June' – rolled off the production line at Castle Bromwich. Test pilot Alex Henshaw drove up from Southampton and took to the air in the first Birmingham Spitfire. He looped and rolled through the clouds to the delight of the assembled dignitaries. For the workers of Castle Bromwich, it was their moment of redemption.

Back in Southampton, though, Denis Webb knew the truth about the 'Ten in June':

> We had, at that time, a contract for Spitfires for Turkey. Eleven had been test flown and sent for crating and shipment. Then on instructions from the Air Ministry the shipment was cancelled and we were instructed to return them to our production line,

and re-erect. We were thus in a position to pass ten complete Spitfires to Castle Bromwich in May, to reappear as the 'Ten in June!' [15]

This was classic Beaverbrook. Appearance beat reality every time. But, on this occasion, it worked. The deception galvanised the exhausted and chastised workforce, and gave them something beautiful to be proud of.

There was just one remaining problem, and it was quite a big one. The largest aircraft factory in the country was a prime target for German bombs.

On 9 August 1940, a single plane dropped its bombs on Erdington, a suburb close to the factory. It was the start of a three-year-long bombing campaign against Birmingham's armaments industry. Just four days later German raiders hit the factory. Eleven bombs found their target and six workers were killed. [16]

Raids on the city and the factory intensified. The workers interviewed by the BBC in 1976 vividly recalled one direct hit:

I got there about half an hour after the bombing and I remember treading on people, there were people blown up in the roof.

And they were stood outside the gate and they said you can come in if you like; the DAs are all over the place.

DAs – Delayed Action bombs – were lying among the wreckage, ready to explode at any moment. But the job of making Spitfires was too important to be stopped. Work simply carried on around them:

I come over the top in my shop in E block, and when
I looked in, the Governor gave me a screw bin and a
piece of mutton cloth and said stick that on your head
and get on with yer job.[17]

Over the years, there's been so much confusion and
misinformation about Castle Bromwich that it can be
hard to get a clear and honest view. For some, it was
nothing short of a national disgrace, with incompetent
leadership and workshy staff who let Spitfire parts gather
dust on factory floors when they were needed most. Why
could this state-of-the-art factory not provide the Spitfires
for the Battle of Britain?

Cutting through the chorus of blame and accusation,
there was one knowledgeable voice of dissent – Supermarine Spare Parts Manager, fixer, mover and shaker
Denis Webb. As a manager, he had access to more information than Southampton's shop-floor youngsters, and he
was a little more cautious in apportioning blame:

When we at the parent firm tendered for our first
production contract for Spitfires, our Chairman Sir
Robert McLean promised that we would deliver our
first aircraft in fifteen months from date of contract.
We thought fifteen months was a reasonable time, to
jig and tool up, and get the first aircraft out.

The contract ordering 1,000 Spitfires from Castle
Bromwich was not placed until April 1939. So, on
the basis of our own judgement, we would not have
expected Nuffield to produce his first Spitfire until
July 1940. And yet in early May 1940 Beaverbrook
was asking Lord Nuffield why he had not produced
any Spitfires![18]

Perhaps history has been a little too quick to judge the Castle Bromwich fiasco. After all, there had never been anything like it in Britain; for scale, for advanced machinery, for sheer ambition. Webb's figures suggest that Castle Bromwich began genuine production very soon after Beaverbrook's 'Ten in June' stunt with twenty-three Spitfires produced in July 1940:

> It seems to me that producing 23 aircraft in 15 months from date of contract was a commendable performance and Lord Nuffield did not deserve such treatment, especially at the hands of a newspaper tycoon who knew nothing of aircraft engineering.[19]

Webb says that thirty-seven followed in August and fifty-six in September. At peak production, the factory was turning out at least 320 Spitfires every month and, by January 1946, Castle Bromwich Aircraft Factory had built 11,694 Spitfires. This was not a record to be ashamed of.

Assistant General Manager Stan Woodley agreed. He'd been seconded to Birmingham from Supermarine, but he was won over by the workers – that much-maligned gaggle of aforementioned car workers, lion tamers and footballers – and he came to respect their commitment to the task: 'All of them did a hell of a job and in the end made a hell of a lot of Spitfires. In one period we were doing 400 Spitfires a month. That was a lot of aeroplanes.'[20]

*

Castle Bromwich Aircraft Factory, after a rocky start, went on to become the most successful Shadow Factory of the war. It set a new standard for production-line engineering, and built over half of all Britain's Spitfires.

The production of Spitfires had been dispersed success-fully throughout southern England and into the Midlands and the Spitfire Fund was helping ease the pressure on the Treasury. The bare minimum number of Spitfires could now be delivered and there was every chance – if Hitler's forces could be contained on the other side of the English Channel – that production could gradually be ramped up to meet the insatiable demand.

But more of the same wouldn't be good enough. Ger-many's talented aircraft engineers hadn't been idle as their bombers pummelled Britain. Lessons had been learned, pre-war designs had been upgraded and brand new fighters were entering production.

The future of the Spitfire – and the security of Britain – depended upon the success of the most important dis-persal of all: the flight of Supermarine's engineers and designers from the dangers of Southampton to an idyllic country estate deep in the Hampshire countryside.

6

The Inspiration of Hursley

Supermarine's design department suffered a direct hit in the September 1940 bombings of the Southampton factory – unexploded bombs ripped through their floor of the office block. They were moved out to the local University College, but within weeks the Luftwaffe had launched its bombing blitz on Southampton. Hundreds of civilians died and an incendiary bomb hit one of the huts occupied by Supermarine. To keep improving the Spitfire, to keep it ahead of Germany's fighter developments, the design team had to get out of Southampton.

As their city burned, hundreds of engineers, draughtsmen and -women, craftsmen and designers moved ten miles north to a very different world.

Hursley House is an elegant Queen Anne mansion set in hundreds of acres of rolling countryside, with stunning views all the way to the English Channel and the Isle of Wight. The Requisition team at the Ministry for Aircraft Production really had delivered for Supermarine. The designers had swapped a ravaged Southampton dockside for a truly inspirational view.

Hursley was long accustomed to playing a role in shaping Britain's history, having once been home to Oliver Cromwell's son Richard. Marrying into the family of

94

the estate's owners, Richard led a rather quiet life until his father's death forced him to assume the position of national leader, as the Lord Protector of the Commonwealth of England, Scotland and Ireland. It wasn't a role that Richard craved or one to which he was in any way suited. His reign was short and chaotic. After just a few months he was overthrown by the army and, on the restoration of the monarchy, he was forced to flee into exile under an assumed name. Once the Royalist lust for vengeance had been satisfied, Cromwell returned home. His body is interred in a vault at All Saints parish church in Hursley.

The estate's original Tudor lodge made way in 1721 for a grand home for Sir William Heathcote, a successful merchant and Tory MP for Southampton. By 1940 it was the home of an American railway heiress, the recently widowed and extremely wealthy Lady Mary Cooper.

The United States may have been reluctant to enter the war, but Lady Cooper was determined to do her bit for her adopted country. During the First World War the Coopers had been one of the largest private financial supporters of the war effort. They had opened their parkland to tens of thousands of soldiers embarking for France and Lady Cooper had converted part of Hursley House into a hospital for army officers, a hospital her eldest son was forced to use when injured in service with the Royal Scots Greys.

With the fear of more casualties in the new war, Lady Cooper again offered her home as a hospital. The requisition team, however, had other ideas. Hursley was far enough from Southampton to escape the bombs but close enough for easy communication with the new workshops there. Despite her surprise at the sudden seizure of her home, Lady Cooper had no intention of holding a grudge. When the first of the Supermarine staff arrived on 7 December 1940, she welcomed them with an enormous floral display – a model Spitfire made out of chrysanthemums.

Sculptures, paintings and Louis XV chinoiserie screens were hurriedly removed to storage, and exquisite seventeenth-century carved-wood panelling and gilded plasterwork was covered with hardboard, but there was no disguising the magnificence of the house and the estate from workers used to the cramped terraces of central Southampton.

The ballroom and adjoining glass-domed winter gardens, saturated with natural light, were ideal for technical drawing, so they were soon crammed with draughtsmen's boards. The Supermarine Technical Office took over the billiard room and the Chief Draughtsman slipped contentedly into Lady Cooper's boudoir. Luckiest of all were the tracers, given the stunning morning room, with its Art Deco lamps, Robert Adam-style Wedgwood plaster reliefs, gorgeous views and cunningly positioned mirrors that still today create an illusion of the room expanding into infinity.

Lady Cooper, along with her youngest son and her domestic staff, remained in occupation. She had her quarters on the first floor while the servants were squeezed into the attic. The kitchens were shared between the family cook and the Supermarine catering staff, churning out hundreds of meals a day for hungry workers.

As more and more staff were transferred from South-ampton, tensions began to rise. Ugly temporary buildings went up, pock-marking the parkland. Close to the kitchen gardens there was a new hangar for the Drawing Office, covered in camouflage netting, and, beside the local school, another hangar for prototype work was being built.

Lady Cooper was used to a clear distinction between family and staff. Workers should use the rear stairs, not loiter in their muddy boots on her grand, polished mahog-any staircase.

Alastair Cooper, Lady Mary's youngest son, was an enthusiastic drinker, always eager for company. Among Hursley's many night workers were the fire-watchers – there to ensure that Supermarine didn't lose a third home to bombs or fire. Alastair made a nightly pilgrimage to the roof, where his crates of brown ale went down well with the bored fire-watchers. Supermarine management were not amused.

Joe Smith, the Supermarine Chief Designer, occupied Sir George Cooper's grand old office. It would have been an ideal space for his many confidential meetings if it weren't for Lady Mary's butler. When he was in need of petty cash to pay the butcher's boy or the window cleaner, the butler would walk straight past Joe and his diagrams of the latest Spitfires to the huge safe built into the corner of the office.[1]

Supermarine management needed an excuse, any excuse, to permanently remove Lady Cooper and her staff. Their chance eventually arrived when a casual kitchen labourer stole some of the family silver and set a fire to hide his escape. The would-be thief was overcome by the smoke and the fire was swiftly put out, but, for the

Cooper family, the damage was done. Unvetted domestic staff couldn't be allowed to work among secret plans for new warplanes.

In December 1942, a small, poignant notice appeared in the Court Circular of *The Times*, tucked just below news of a visit to Buckingham Palace by the King and Queen of the Hellenes: 'The Dowager Lady Cooper has left Hursley Park, Winchester and is now residing permanently at Jermyns, Romsey.'[2]

At last the Spitfire designers had the place to themselves and plenty of space to recruit a new generation of design talent. Through the gates of Hursley Park cycled a very determined eighteen-year-old art student from Portsmouth by the name of Stella Broughton:

> On my first day at Supermarine, I left home about 6 a.m. . . . The cycle was some thirty odd miles . . . After having packed my bike with all I needed for the next week I started to climb up Portsdown Hill – a climb of six in one! Past Fort Widley, turned down Pigeon House Lane. For three miles there was no need to pedal.[3]

On leaving college, Stella had found work as a tracer at HMS Excellent, a naval station where she transferred designs for warship parts on to tracing paper for reproduction. It was important but rather dull work. Stella was hungry for a new experience:

> When I became bored with my job as a tracer in the experimental Drawing Office at Whale Island, I enquired if there was any way in which I could improve my position, and the reply was, as a girl

under 21 years and with no HND qualification, there was no possibility of anything else being available.[4]

Stella wasn't the sort of girl to take no for an answer. She'd get what she wanted even if it required the application of a little light nepotism. Her father, a tutor at the Southern College of Art, bumped into Gerald Gingell in Southampton. Gingell was head of Supermarine's technical publications department and he was on the lookout for an artist who could read plans and draw freehand in perspective. Did Mr Broughton know of anybody who might fit the bill?

> I was literally commandeered from working for the Navy to go and work on the Spitfire, because my skills were required and no one else had these particular skills.[5]

There were plenty of young women working at Hursley, but Stella joined a section dominated by men:

> The Drawing Office comprised some 200 draughtsmen, and my appointment had caused some embarrassment at the fact I would be the first female of under 21 years to sit at a board.[6]

Stella, with her infectious laugh and boundless self-confidence, certainly stood out. She found some of the older men dismissive of what they called 'pretty pictures', but Stella wouldn't let spite or jealousy hold her back. She knew she was doing vital work for the war effort:

> We had to clock in and out every day, then I'd get to my desk where I'd join my colleagues, three

draughtsmen who prepared the initial drawings for the assembly of parts of the aircraft. Those drawings were given to me and I had to do the master linen copy of these drawings for printing. Those were then sent out and put into crates which contained parts for the assembly of the aircraft.[7]

Stella was told that the crates were being sent to Lieutenant-General Montgomery in North Africa. Her drawings guided the RAF engineers supporting the Desert Rats in the assembly of Spitfire parts. In the war in the Western Desert, modern weaponry was in short supply – every Spitfire really did count.

Maintaining the quality of her draughtsmanship was hugely important to Stella – nothing would stop her producing the best possible work for the fighting men and the engineers and ground crew who supported them:

At college I was top in lettering and it was the lettering skills that were very important to Supermarine because all the written stuff on the masters had to be read by people who didn't really understand English. So it had to be very, very clear.[8]

Young Stella Broughton was a perfectionist, just like the late father of the Spitfire himself. The spirit of R. J. Mitchell, the stubborn, temperamental engineer from Stoke-on-Trent, drove the team at Hursley Park. It was his blend of engineering know-how and artistic genius that had given birth to the Spitfire and it was his inspirational leadership that bonded the team now stretching every sinew to make his creation ever faster and feistier. If they failed, Britain could lose her mastery of the skies.

To understand why the construction and development of the Spitfire became so important to the defence of Britain, we need to spool back to a more innocent time in the early days of aviation. It was a time when speed was an international obsession and an intense young English engineer stood on the brink of greatness.

Part Two

THE DESIGNERS

7

The Father of the Spitfire

Baltimore, USA, 23 October 1925: the Supermarine S4 is a sensation. At a moment in aviation history when the RAF's top fighter is a wooden biplane that struggles to reach 140 mph, the S4 seems to have beamed in from another planet. A sleek, silver rocket set on top of two huge floats, this revolutionary monoplane has already flown faster than any British aircraft before, hitting 226 mph. *Flight* magazine describes the design as 'inspired' and 'courageous'. Its young designer has seized the attention of the aviation industry and he has every reason to be proud of his beautiful creation, but today, R. J. Mitchell is pacing the shore of Chesapeake Bay, staring at his feet, deeply concerned.

Mitchell has spent the last few days living in a tented village a few miles outside Baltimore. This is the British team base for their attempt to win the Schneider Trophy, a hard-fought annual competition for the fastest seaplanes in the world. It's a race that's driving forward the technology of aeronautics at a breathless pace, but at a terrible cost in pilots' lives. The S4's top speed is higher than its Italian, American and British competitors, but preparations for the race have been far from ideal. The weather has been abysmal – gales have made practice runs impossible and brought down a heavy tent pole on the tail section of the S4, requiring complex repair work. Worse still, Henri Biard, the only man Mitchell trusts to fly this complex marvel, sprained his ankle on the Atlantic voyage and then caught influenza on arrival in the United States.

Biard insists he's fully recovered. He squeezes into the open cockpit for the first of the navigation tests that must be passed before the competition begins in earnest. As Biard taxis the S4 across the Bay, Mitchell is short of confidence in his plane and his pilot. The take-off is smooth, Biard turns back to shore and circles the tents of the British team. At a height of about 250 metres, he gains speed, crosses the pier and makes a sharp turn. Excited spectators cheer what looks like the start of an aerobatic manoeuvre but the S4 seems to stall. It slips right and then left, apparently out of control before twisting tightly like a falling leaf, hitting the waves with a heart-stopping crash. The S4 catapults violently on to its back. Captain Biard is surely dead. Mitchell jumps into the high-speed rescue launch that races him across the bay. The S4 has failed disastrously and the reputation of R. J. Mitchell hangs in the balance.

*

Reginald Joseph Mitchell grew up in the adolescence of aviation. He was a fascinated witness and then a prime player in the thirty years of high-octane development that took humanity from the ground to the air, from the 20 mph Wright Flyer skimming the sands of Kitty Hawk to his own designs, which could reach 400 mph and fly nine kilometres high. Born in Staffordshire in 1895, he was the son of a Yorkshire headmaster who went on to run a successful printing business. In the family's large Stoke-on-Trent garden, Reg and his two brothers and two sisters kept racing pigeons and built model aircraft from bamboo and rubber bands, inspired by the aviation craze sweeping the country. One of the Wright brothers' early aircraft was displayed in Stoke's Hanley Park and the *Daily Mail*'s London to Manchester air races kept young Reg's eyes to the skies.

At Hanley High School, Reg excelled at Maths and Art. In 1911 he was apprenticed to the local railway locomotive engineering firm Kerr, Stuart and Company. For a solidly respectable middle-class chap with his head somewhat in the clouds, the grease, grime and hierarchy of the engineering works came as something of a shock. He complained to his father about having to travel to and from work in filthy overalls and he frequently clashed heads with his superiors. Reg particularly disliked making tea for the crew. When his foreman complained loudly that the tea 'tasted like piss', Reg is said to have brewed up the next pot with his own urine. R. J. Mitchell would never be one to follow orders, but he always knew how best to deploy his rather cruel sense of humour. His fellow workers apparently found it hilarious.

By the time Mitchell finished his apprenticeship, the First World War was in full swing and the early aeroplanes were proving their value as artillery spotters over the trenches of France and Belgium. He tried to join the army but was told that his engineering skills were too valuable to be wasted in the forces. Instead he answered an advertisement for a job as personal assistant to the owner of a small aircraft-building firm in Southampton that specialised in seaplanes.

Supermarine was a rather odd little company. It had been founded in 1913 by Noel Pemberton Billing, an aviator, inventor and politician who harboured a dangerous obsession with homosexuality. He was particularly concerned by the supposedly evil influence of lesbians in public life.

Billing had run away from his Hampstead home to South Africa at the age of thirteen, making his way as an actor, boxer and policeman, before returning to England, where, like many dreamers of the time, he developed a fascination with aviation. He raised enough money to rent a large shed by the River Itchen in Southampton and set about turning his ideas for flying boats into reality. The 1914 declaration of war with Germany came at the perfect time, prompting the government to order new designs from companies large and small. Billing had plenty of ideas to offer and one of those – a Zeppelin interceptor, dramatically named *The Night Hawk* – would be the first aircraft that R. J. Mitchell would work on.

By 1916, Billing was so horrified at the incompetent handling of the war that he stood for Parliament, campaigning for the creation of a British air force, separate from the command structures of the army and navy. As the newly elected independent MP for Hertford, he demanded air raids on German cities and radical improvements in the aircraft being sent to the Western Front. To spread the influence of his anti-establishment ideas, he set up a magazine, *The Imperialist*, later ominously retitled as *The Vigilante*. The magazine soon became a haven for deranged racist conspiracy theorists, with contributors railing against the draining influence of Jews and German music. More specifically, the magazine fixed on rumours

of an establishment homosexual and paedophilic conspiracy. It claimed 47,000 prominent Britons were being
blackmailed by the German government because of the
pleasures they took in 'evils which all decent men thought
had perished in Sodom and Lesbia'. Their names were
said to be contained in a 'Berlin Black Book'. There were
even strong hints that the Prime Minister's wife, Margot
Asquith, was caught up in this dastardly conspiracy.

*

The story of the first decades of the aviation industry is
packed with moustache-twirling eccentrics, daredevils and
mavericks. Reg Mitchell's job application had brought him
right into the heart of it. His rather conventional Midlands
upbringing may not seem, at first glance, to have equipped
him for this world of chancers and minor aristocracy,
but buried beneath his solid engineering qualifications, a
slight stammer and a superficially shy manner, beat the
heart – and temper – of an artist. He seems to have been
able to charm wealthy investors just as readily as he could
tame a shed full of railway mechanics.

Supermarine's product line at the time Mitchell joined
was very much cast in the image of its eccentric founder.
Billing had become incensed at the ease with which
German Zeppelins had bombed London and the east
coast of England. He responded with a series of designs
for fighter aircraft that could fly sufficiently high to reach
the airships and carry weapons capable of shooting them
down. The Night Hawk was a quadruplane with four
pairs of wings stacked on top of each other, a powerful
moveable searchlight at the front, a Lewis machine gun
in the nose and a cannon mounted on the top wing.

Fortunately for its prospective crew, the war was coming to an end by the time this baroque wedding cake of an aircraft was ready for service.

During the war, Billing had sold the company to his friend and factory manager, Hubert Scott-Paine. A large, gregarious redhead with a passion for speed, he was no more a conventional industrialist than Billing, but Scott-Paine clearly appreciated the ambition and talent of the company's new recruit and, within months, Mitchell was promoted to Assistant Works Manager. The extra wages that came with the new position gave Mitchell the confidence to return to Stoke and marry his girlfriend, Florence Dyson, headmistress of an infant school.

Eleven years senior to R. J., Florence is credited by biographers with providing the domestic stability that left him free to focus his attention on each of the two dozen different aircraft designs he would complete at Supermarine. Judging by reports of his short temper and his obsessive behaviour, Mitchell cannot have been an easy man to live with, but Florence doesn't seem to have felt neglected. She would go on to play a pivotal role in the beatification of Mitchell, helping the producers of the much-romanticised biopic, *The First of the Few*. An interview published during the Battle of Britain suggests that she was content to enjoy the acclaim, and the interesting characters, that her husband's growing reputation brought to their Southampton home.

The first months of peace gave Mitchell the breathing space to learn his new trade. Textbooks on aeronautics were now being published and devoured by a new generation of young engineers. The days of building 'by guess and by God' were fast disappearing. This was the perfect time for a polymath like Mitchell to shine – a trained engineer, but one who could lift his head from the drawing board and imagine something better, something faster, something beautiful.

Mitchell's radical imagination wasn't called on immediately. Supermarine was busy with contracts to convert military surplus flying boats into civilian planes. Some of the flying boats that Mitchell worked on were sold on to the Norwegian Navy, while others took part in a photographic survey of the vast Orinoco River delta in Venezuela. Aeroplanes were just beginning to open up the world, but the future for Mitchell and Supermarine would be determined less by this new age of discovery than by the relentless march back to war.

*

The First World War had turned British aircraft production from a shed-based hobby to a full-scale industry, but peace slashed the state aviation budget. Innovation ground to a halt. Around two dozen small-to-medium-sized companies were now scrabbling desperately to secure orders. Supermarine certainly wasn't one of the biggest or the most influential – any government orders would be very welcome.

David Lloyd George's post-war coalition government was caught between two competing forces. The popular desire for peace after five years of death and destruction was overwhelming and the coffers were largely bare. Britain simply couldn't afford the latest weaponry.

On the other hand, Britain still had an enormous empire to maintain. The war and the subsequent peace treaties had given birth to new nations and injected fresh hope into independence movements around the globe. If the British Empire was to remain intact, then aviation surely had a role to play in communications and colonial warfare. The Royal Air Force had certainly not been idle since the Armistice. British planes had fought Bolshevik forces in Russia and RAF bombers had seen extensive action against revolting tribes in Mesopotamia. As part of the Versailles Treaty, the Middle Eastern territories of the collapsed Ottoman Empire were shared among the victors. These 'mandates' were supposed to be kept at peace by the British and French until native governments could be established. In many areas, the ever so subtle distinction between a colony and a mandate escaped the local people – they saw that one imperial overlord had simply been replaced by another.

In the area now covered by Iraq, Sunni and Shia tribes united against the British. It took a reported 100,000 troops to pacify the oil-rich region. What did not escape the beady eye of the War Secretary, Winston Churchill, was the role played by the RAF in bombing Mesopotamian villages. The shock and awe provoked by attacks on civilians from the air pacified large swathes of territory. Two lessons could be drawn from this – the first and most stubbornly persistent was that bombing civilians could win a war. The second, slightly more nuanced message that Churchill drew from Iraq was that the empire could be controlled more cheaply by 'aerial policing' than enormous standing armies. This new policy of 'control without occupation' would need new kinds of aircraft. Those planes would have to be designed and built by a healthy

British aeronautics industry, so the government set about spreading seed money through the industry to encourage innovation.[1]

Supermarine, with its pre-war specialism in wooden flying boats, was well-placed to take advantage of these new demands. Aircraft that could carry colonial officials, soldiers or bombs to any corner of the Empire without the need for airstrips or complex repair facilities appealed to the government. R. J. Mitchell set to work designing a series of flying boats, large and small, to serve a wide variety of purposes. There was the passenger-carrying Commercial Amphibian, as well as a small spotter plane for the Navy – the Seal II – and the Sea King range of fighters, designed to be launched from the new aircraft carriers. The Spanish government bought half a dozen of his Scarab flying boat bombers to use in their long-running and increasingly vicious Rif war against Moroccan forces fighting for independence.

Fresh unrest in Egypt underlined the need for fast communication links between London and the great imperial cities. The Air Ministry had ordered six airships to maintain an air-bridge, but slow speeds and technical problems convinced civil servants that flying boats would provide a more reliable service to the Near East. Officials had been so impressed by Mitchell's early designs that they ordered his latest and largest aircraft yet – the Southampton – straight from the drawing board. Seven of these rather stately planes were ordered in 1924, a hefty financial commitment for the time. With a range of 500 miles and an ability to fly on just one of its two engines, the Southampton was a practical patrol plane for the furthest reaches of the empire. Equipped with hammocks, a basic stove and even a toilet of sorts, it fast became a favourite aircraft of pilots and crew. Sales of the Southampton were made to the navies of Japan, Argentina and Denmark, while civilian airliner versions were flown by Imperial Airways (forerunner of British Airways) and Japan Air Transport.

The Southampton proved to be a useful money-spinner for Supermarine, but it wasn't to be these hefty flying boats that would make Mitchell's name and it wasn't them that would inspire the Spitfire. For that story we need to turn to the extraordinary planes Mitchell built to win the Schneider Trophy.

8

The Schneider Boost

Hubert Scott-Paine could see the difference that Mitchell was making to his small company on the banks of the River Itchen. New designs seemed to flow from his pencil and his work inspired confidence in the staff, in pilots and, most importantly, in the influential buyers at the Air Ministry. Mitchell was appointed Chief Designer of the company in 1919. He was just twenty-four years old. In the aviation industry this was a time for youth. Well educated in mathematics, but free of the iron-clad precon-ceptions of Victorian heavy engineering, Mitchell and his rivals at A.V. Roe and Vickers were all in their twenties. As with the technology entrepreneurs of today, these young men were well rewarded for the calculated risks they took to get brand new aircraft from the drawing board to the airstrip in a matter of months.

To industry insiders, R. J. Mitchell was already a name to drop, but heavy flying boats and naval patrol craft wouldn't set the public pulse racing. In the 1920s, after years of war, there was a yearning for distraction. Speed and glamour were in demand and it was Mitchell, the brooding, sandy-haired engineer from Staffordshire, who would soon grasp his opportunity to design the world's fastest and most beautiful aircraft.

It was the personal obsessions of one wealthy man that gave Mitchell the chance to show what he was capable of. Jacques Schneider, Parisian heir to an armaments company, became fascinated by aviation after seeing Wilbur Wright demonstrate his latest Flyer at Le Mans in 1908. Schneider had a hunger for speed, which was only sharpened by a hydroplane boating accident at Monte Carlo that left him permanently disabled. A record-breaking balloonist and keen amateur pilot, he became frustrated with the slow progress of the pre-war civilian aircraft industry and so offered a handsome prize to the engineers who could create the fastest planes on the planet. As far as Schneider was concerned, that meant seaplanes. Aircraft needed lots of space to take off and land and most of the major cities of the world were built by rivers, lakes or open sea. Schneider had a vision of the future: high-speed inter-city seaplanes would transport international travellers and business people directly to their final destination.

The Schneider Trophy, launched in 1913, caught the imagination of aircraft builders, who spotted an opportunity to test out their latest ideas against the very best opposition, and gain valuable publicity in the process. After the First World War, national governments were alert to the military value of high-speed aircraft and the trophy proved to be catnip to the new popular press of the time. The combination of power, speed and death-defying aeronautics offered a free show full of glamorous characters that the newspapers revelled in, turning the top pilots into international pin-ups. At a time when horses and carts still ruled most European streets, the sight of 300 mph aircraft racing overhead was hugely exciting. An estimated half a million people watched the 1931 race from the shores around Southampton.

As the owner of a small company that specialised in seaplanes, Hubert Scott-Paine couldn't resist the challenge of the Schneider Trophy. It offered an incredible opportunity to showcase Supermarine technology. From 1919 to 1931, R. J. Mitchell and his team entered a series of ever-faster aircraft into a race that developed into a timed circuit of around 200 miles on a pre-arranged course.

For a flavour of the 1919 race, it's perhaps worth watching the 1965 comedy of early aviation, *Those Magnificent Men in Their Flying Machines*. Mitchell's first Schneider experience proved to be a very British farce. Supermarine entered its Sea Lion, a flying boat designed by Mitchell's predecessor, F. J. Hargreaves. As the Sea Lion lined up alongside the other five competing aircraft from France, Italy and Britain at Cowes, on the Isle of Wight, a blanket of fog descended. The bustle of mechanics and the roar of the engines was smothered. Eventually, impatient French and Italian flyers attempted a circuit of the Solent, but when they landed their aircraft were damaged by over-enthusiastic bathers sculling out from Bournemouth beach. The British competitors then took to the air, but the entries from Fairey and Sopwith hastily turned back when they hit a fresh bank of fog. The pilot of the Sea Lion took off, got lost, landed to work out where on earth he was and up-ended the plane, tipping himself into the water. One of the Italian aviators successfully completed the ten laps of the circuit but he was disqualified when found to have taken an accidental short cut. The 1919 Schneider Trophy was declared null and void, much to the embarrassment of the British organisers.[1]

Mitchell's first attempt as a designer was in 1922. Italy, now the acknowledged leader in high-speed seaplanes, had won the 1920 and '21 races held in Venice. Under Jacques Schneider's rules, a third victory would hand them the trophy permanently. That was a possibility that Scott-Paine, with his pride in British engineering, would simply not stand for. As a businessman, he was also keen to gain publicity for Mitchell's latest design, a small ship-borne fighter that had stubbornly failed to sell. The Sea Lion II was a clumsy-looking amphibious biplane with a propeller that pushed rather than pulled the aircraft. Powered with a Napier Lion engine, it was capable of around 145 mph. This was theoretically fast enough to beat the Italians, but, as Supermarine test pilot Henri Biard told the BBC in 1957, it was a difficult and dangerous plane to fly:

> The only trouble with these things was the take off. That was the most dangerous thing of all. You had the design to give its maximum efficiency at high speed. When you started over the water, if you hit a swell from a motorboat or any kind of thing you'd just bounce into the air. And it wouldn't fly, it used to come down and go wallop! And then up again. And I can't tell you how many people were killed that way.[2]

Henri Biard was selected to pilot the Sea Lion II around the Bay of Naples in the 1922 Schneider Trophy race. On arrival in Italy, he discovered that the local engineers had heard the rumours about his plane. They appeared to be rather concerned by their over-powered, unstable rival:

When we ran the engine on the ground, we had a lot of vibration. And one of them (Italian engineers) said to me, 'Oh, I think the wings will collapse in the air'. They spoke English, funnily enough. And I said, 'Well, sailors don't care'.

Captain Biard was no sailor, but neither was he a man to duck a challenge. Born in Surrey to a French father and English mother, he had, in the early days of the war, witnessed German cavalry burning down his grandfather's farm. He'd sprinted away from the advancing Germans and made it back to Britain to help train the first generation of fighter pilots.

On a stifling hot and humid day in Naples, Biard took his first test flight dressed in his regulation flannel shirt and trousers. Treating the flight as a tourist excursion, he diverted over Mount Vesuvius. The thermals rising from the crater caught him unawares and the light little plane was blown miles off-course. If his stiff upper lip had wavered, Biard certainly wasn't going to show it in front of those skittish Italians. He landed with exquisite nonchalance and began his preparations for the race itself, 200 miles round and round the gorgeous turquoise sweep of the Bay of Naples:

When the time came, the Italianos went first. I was just getting into mind to go after them when Scott-Paine came along, and gave me a bottle of brandy to shove in my pocket.

Suitably fortified with much-needed Dutch courage, sweating profusely in his flannels, Biard took off into the bright blue Neapolitan sky and chased after the two Italian aircraft:

It was very, very hot and very bumpy too. On every corner I came at, there were the two Italianos in front of me to stop me passing them. And their high speed machine was going on to win the thing.

This was a time trial, not a race, but Biard was oblivious to the rules, caught up in the thrill of the chase:

Finally I couldn't stand this thing so I got above them. My dive just over them nearly blew one upside down. And I came up at the end for the front place.

With the Italians left scattered and fuming in his wake, Captain Biard crossed the finish line:

I didn't know if I'd won it, of course, and I turned back over the place and did a loop, which I thought would impress them.

Biard's free-form aerobatics proved the last straw for his highly-tuned Napier Lion engine:

Funnily enough, after that the engine started to fall to bits. One of the pipes burst – the water was flying all over the place – but it was only four miles and I shut the engine off. I was high enough I could glide it back.

On landing safely, he was met by the beaten Italian pilots:

They were very nice, awfully nice. They said I'd won it. I suddenly remembered this flask of brandy, which

I had in my pocket and I said, 'I have brandy', and we all drank it.

Not all of the Italian crew were quite as magnanimous in defeat as the pilots. Supermarine owner Hubert Scott-Paine had endured the crowd's reaction to Biard's aggressive tactics and his shameless show-boating:

> They're not happy that you're doing this thing, looping over them. I think it was the first flying boat to loop the loop. The next morning, we were walking on the front where we saw the commanding officer of this small Naval Air Station who had been very nice to us. And we said to him, 'Good morning'. He looked at us and he spat on the ground and said, 'Bloody English!'[3]

The welcome back home to Southampton was somewhat warmer. The Mayor met Biard and Scott-Paine at the railway station and they led a procession to the Floating Bridge – the ferry that linked the city centre to the Supermarine factory in Woolston. Henri Biard, a tall, striking figure with the chat to charm a crowd, made the perfect hero for the town's schoolchildren. Fast-growing Southampton could now look to the skies for its future prosperity, not just to the ocean.

Excitement mounted ahead of the 1923 competition. It would be held at Cowes on the Isle of Wight, just a few miles across the water from Supermarine's Woolston home. The eager new audience for high-speed flight would have a spectacle of speed right on their doorstep. With an almost empty order book, Scott-Paine didn't feel he could justify the design and construction of a brand new aircraft so Mitchell was handed back the Sea Lion II and told to do what he could to upgrade it. He improved its aerodynamic performance with a new set of wings and a smoother hull. The Napier Lion engine was tweaked to increase its power to 550 hp. In theory this should push the newly numbered Sea Lion III to 160 mph.

Mitchell was always racked with tension as race day approached. Would his plane perform to its potential, would his pilot make it safely home? His mood could not have been improved when the rival national teams sailed into harbour. Capturing the excited attention of the city, the newspapers and flying fans across the country was the team from the United States. Sponsored by the US Navy, this team had the money, the teeth and the technology. In beautifully pressed naval uniforms, the handsome pilots – Rittenhouse, Irvine, Gorton and Wead – cut an undeniable dash as they slid into their elegant new aircraft. These planes – two Curtiss CR-35s and a Navy-Wright NW-2 – were nothing like the flying boats of previous competitions, they were high-speed biplanes with twin floats attached. Older designs seemed to defy their boat-builder body shape to reach high speeds; these American machines looked born to race.

The one hope for Mitchell and Supermarine was the competition's notorious rate of attrition. With a series of tricky navigation and mooring tests to pass before race day, there was always the chance that hulls would leak or temperamental engines explode. The competitive field might just thin out in Supermarine's favour. Sure enough, the other British entry, the Blackburn Pellet, overturned on the Solent, the Navy-Wright crashed and the engines of two French competitors failed. That left one Frenchman, two Curtiss CR-3s and the Sea Lion. Captain Biard flew with all the speed and cunning he'd demonstrated in Naples, but there was no chance of catching the Americans. Both Rittenhouse and Irvine outflew Biard by 20 mph and took the Schneider Trophy back across the Atlantic.

Mitchell was left to ponder the lessons learned. The flying boat – essentially a boat hull with wings and engine attached – had reached its limits. The Sea Lion III had a more powerful engine than the US Navy aircraft but horsepower alone clearly wasn't enough. To win the trophy back would need a completely new design, not just a new engine. He somehow had to leapfrog the technological advances of the Americans. Supermarine had complete confidence in his design skills, but Mitchell would need time and money to come up with a world-beater.[4]

As the date of the 1924 competition approached, the Air Ministry released funds for Supermarine and the Gloster aviation company to design potential competitors, but Mitchell scrapped his initial ideas and Gloster's first attempt crashed. Britain would not be in a position to compete. Italy and France, too, failed to construct new machines in time, so the Americans, rather sportingly, agreed to delay the race for a year.

For the first time Mitchell had the space and time to follow his instincts. The new aircraft would not just be a single step forward from an existing Supermarine design; it would kick-start an aeronautical revolution. Napier had a powerful new variant of their Lion engine in development. That engine would be contained within the fuselage, not bolted on top. The new plane must be a monoplane to ensure maximum aerodynamic efficiency. Mitchell was convinced he could build a pair of wings strong and stable enough to avoid the need for the bracing wires and struts that reduced the aerodynamic efficiency of contemporary monoplanes.

The result looked like nothing built before. A thin, streamlined fuselage and an enormous two-blade propeller sat delicately on top of a huge pair of floats. The lines were clean and elegant. There were no wires, no flapping sheets of canvas, no visible pipes or pistons. There was nothing here that connected this plane to the first age of powered aviation. *Flight* magazine was very impressed:

> One may describe the Supermarine Napier S4 as having been designed in an inspired moment. That the design is bold no one will deny; and the greatest credit is due to R. J. Mitchell for his courage in striking out on entirely new lines.[5]

This stunning creation was built in just five months. Hubert Scott-Paine had sold Supermarine to Commander James Bird, who was content to let his in-house genius lead on all design matters. Mitchell had gathered a small, hand-picked team of designers and draughtsmen around him. Observers described them as 'young and raucous', given to practical jokes and lavish parties, but they were talented, dedicated to the company and ferociously loyal to Mitchell.

His collaborative working method produced innovation at enviable speed. He would walk into the drawing office and study a particular problem on a board. One by one he would pull other designers into the conversation, giving everyone the chance to contribute towards a solution. Unconsidered or foolish thoughts, though, were given short shrift. If he disliked a drawing, it was simply pushed off the desk without a word spoken. It was this team and this creative process – what one staffer described as 'an atmosphere of continual achievement' – that would bring the Spitfire to life and keep Supermarine ahead of the competition right through to 1945.

Captain Biard took the S4 for its maiden flight on 24 August 1925. Biard immediately noted problems with the pilot's view from the cockpit – he came close to hitting a liner on take-off and a dredger on landing. For speed, though, the S4 fitted the bill perfectly. Within weeks Biard had travelled faster than any Briton before, reaching 226.75 mph, fully 70 mph faster than Supermarine's previous Schneider entry.

In mid-September the S4 was loaded carefully on board the SS *Minnewaska*, bound for New York. The Supermarine crew was photographed on deckchairs in mid-Atlantic. All in regulation tweed suits and ties, Biard seems the most relaxed, shading his face from the sun. Mitchell chews on his pipe and glares at the photographer. For Mitchell, the unbearable pre-race tension was already ratcheting tight. The aircraft packed in the ship's hold could fly at speeds man had never reached before. The S4 was built for speed and nothing else. The life of the man sitting next to him depended entirely on Mitchell's calculations.

The other entrants seemed beatable. The British rival, the Gloster III, looked like a direct copy of the US 1923 winner, while the Americans themselves had simply uprated that triumphant Curtis CR-3. The Italians had switched to a monoplane format like Mitchell, but the body of the aircraft was a traditional flying boat hull with an external engine.

The Supermarine team was the first to arrive on the shores of Chesapeake Bay, but illness and weather delayed their practice flights. Eventually, on 23 October, all the competing aircraft were in position and able to begin their navigation test flights. One of the Americans took off first, followed closely by a Gloster III and then Biard in the S4. Biard described his short flight to the BBC:

> The S4 took off beautifully until I did a slight turn when this wing wobble started to such an extent, something frightful, and the joystick was in front of me like this – I couldn't hold it. It was going 'zonk', 'zonk'! I couldn't hold it at all.

The wings were vibrating so violently that there was nothing he could do to keep the plane in the air. Henri Biard prepared to meet his Maker, spiralling toward the waves at 200 mph:

> I got a frightful wallop on my head, which has always been my weak spot. Oh, the noise was like something frightful! When I came to I was at the bottom of the sea and the S4 was on top of me. Or what was left of it.

As Biard struggled to free himself, a fast rescue launch sped across Chesapeake Bay, heading for the crash site. It broke down halfway and another boat had to be summoned:

> I had an awful job getting out of it but I did finally get out and rose to the surface. Fortunately I was able to hold my breath for a long time but I almost burst coming up.[6]

Biard was alone in the water for an hour. When rescue finally arrived, R. J. Mitchell was an unwelcome sight in the prow of a boat. He asked Biard, 'Is it warm?' It certainly wasn't, but somehow the frozen and furious Captain Biard had survived with just two broken ribs and a nasty spot of concussion.[7]

Mitchell was distraught at the loss of the trophy and his aircraft, and the damage done to his friend. The faster these planes got, the less chance a pilot had of surviving a malfunction. Without modern diagnostics, the cause of the crash could not be discovered with any certainty. In his detailed study of the Schneider races, the aviation historian, John Shelton notes that the official crash report stated simply that 'the S4 stalled and crashed into the sea'.[8] The correspondent from *The Times*, however, reported wing flutter as the cause.[9] This oscillation of the wings occurred in all aircraft, but the ever increasing airspeeds magnified the problem. Mitchell had perhaps been premature in doing away with the supports, wires and struts that most monoplanes used to maintain their stability. Certainly, Mitchell's next high-speed aircraft would use external wires to stabilise the wings.

The 1925 race continued without the S4, the home team's Curtiss beating the Italian challenge with an average speed of 232.57 mph. The Italians drowned their sorrows in Chianti, supposedly smuggled into Prohibition-era America in the floats of their planes.

*

In 1926, British manufacturers had nothing to offer against the American flyers, but the Italian dictator, Benito Mussolini, ordered two of the country's industrial giants,

Fiat and Macchi Aeronautica, to retrieve the Schneider Trophy and return it to its rightful home. The stunning success of the M39, taking first and third place at the competition held at Hampton Roads, Virginia, alerted the British military to what should already have been obvious. The Trophy was being used as a preparation for war and it was the Italian Fascists who led the world.

A concerted effort was made by the British government for the 1927 competition, with three different aircraft funded by the Air Ministry. The cutting-edge facilities of the Royal Aircraft Establishment were offered to the manufacturers, who made full use of the wind tunnels and water tanks available at the Farnborough site. Noting that rival teams from Italy and the USA had used military pilots, a new Royal Air Force High Speed Flight Unit was set up to help win the trophy and advance British experience at the extreme limits of aviation.[10]

Mitchell toughened up the Supermarine S5, ditching much of the wooden construction of the S4 in favour of a more rigid all-metal fuselage. The wings were strengthened with wire struts to cut down the flutter that may have contributed to the crash of the S4. On its first flight in June 1927, the S5 reached an unofficial speed of 284 mph.

Britain clearly meant business, an impression that was reinforced by naval swagger. Mitchell and the S5 arrived in Venice on board the aircraft carrier, HMS *Eagle*, accompanied by four Royal Navy destroyers. Mussolini may have been the acknowledged 1920s master of chest-beating pomp – Venice was decked in flags and the Schneider Cup was on display in St Mark's Square – but the British, too, knew how to put on a show of national might. British confidence was, on this occasion at least, not misplaced. In the perfect aerial amphitheatre of the Venice Lido, tens of thousands gathered on islands, boats and gondolas to watch Flight Lieutenant Webster of the RAF High Speed Flight win the race with ease, posting a new world speed record of 281.65 mph. The Italians seemed to take their defeat with good grace – their national hero, the playboy, poet and proto-Fascist politician, Gabriele d'Annunzio, presented Webster with a huge ruby-and-gold ring.

The British team returned to another hero's welcome in Southampton. The Mayor toasted Mitchell in champagne, declaring that, 'You have done something for England that will live through generations to come.' Supermarine owner Commander Bird replied that Mitchell had 'out-designed the Italians'.[11]

To Mussolini and his brilliant Italian engineers, that was fighting talk. A few weeks later their elegant seaplane, which had suffered engine failure in Venice – the Macchi M52 – overtook the S5's world speed record. The British team was convinced that, in the right conditions, the S5 could be pushed just a little faster, perhaps even beyond the new Italian record of 297 mph. South African fighter ace Samuel Kinkead took the controls on 11 March 1928. The weather was grim and several appointed start times came and went. Eventually, impatience got the better of the team and Kinkead was cleared for take-off. Little could be seen from the land, but it seems that Kinkead misread his altitude in the poor visibility and crashed straight into the sea.

*

Mitchell had got to know the young pilots of the High Speed Flight well. They would often gather at his Southampton home for drinks and conversation. Mitchell enjoyed their company and he genuinely welcomed their comments on the flying qualities of his creations. These were not Supermarine employees, they were self-confident boys with a public school education – they felt free to speak their mind even to a highly respected designer. Mitchell also saw these evenings as an opportunity to get the measure of their individual personalities. As an engineer he liked to be in control of all the variables of flight. His time spent with each pilot transformed them in his mind from unpredictable to predictable variables. Kinkaid's death hit Mitchell hard. The S5 was a tough plane to fly, but his colleagues told him that there were no structural faults that could have caused the pilot's death. Mitchell, though, knew the truth. High-speed aviation was in its infancy. Every time he asked a pilot to take one of these machines into the air, he was asking them to dice with death. He revealed his true feelings in a speech to a Rotary Club meeting in Southampton:

> The designing of such a machine involved considerable anxiety because everything had been sacrificed to speed. The floats were only just large enough to support the machine, and the wings had been cut down to a size considered just sufficient to ensure a safe landing . . . In fact everything had been so cut down it was dangerous to fly.[12]

Despite the ever-increasing danger of the Schneider competition, the government was determined to hold on to Britain's hard-won trophy. The S5's Napier Lion engine had been consistently uprated since its first use in the 1919 Schneider Trophy, doubling in horsepower in eight years, but it was felt now to have reached its limits, so the Rolls-Royce car and engine company was ordered to come up with a new powerplant for the next race. The directors of Rolls-Royce were more interested in their luxury car business than aero engine development, but arms were twisted and Henry Royce began a close working relationship with R. J. Mitchell that would ultimately prove vital to Britain's war effort.

As Rolls-Royce worked on a bespoke engine, Mitchell further refined his designs, completing the transformation to an all-aluminium construction. Mitchell's son, Gordon, watched the S6 fly and remembered how much these aircraft meant to his father:

> I well remember the anguish and tension which my father showed when the Schneider Trophy planes were wheeled out and the engines revved up. I remember the large floats, the big silver wings. This was all to me for a boy of seven or eight, as I was then, something of very great excitement.[13]

Supermarine staffer Arthur Black confirmed young Gordon's impression of Mitchell's complete focus on the job in hand: 'When early test flights of a new aircraft were in progress his concern was so great that it paid not to attempt polite conversation.'[14]

Rolls-Royce delivered a new engine as promised, but its raw power proved problematic. To reach the required speed, it ran at such intensity that engine parts would begin to fail after less than an hour's running. Even when those teething problems were resolved, each engine would have to be returned after just five hours of flight to the Derby factory for a complete overhaul. The extra power and weight also unbalanced the slim Supermarine fuselage. Pilots found it hard to take off and tricky to control once in the air. What the pilots did appreciate, however, was Mitchell's analysis. He always seemed able to warn them of the particular eccentricities of the prototype aircraft; he could explain how a particular tweak to the loading, the engine or the airframe would impact their flying experience. In common with the Supermarine technicians and Rolls-Royce engineers, the pilots also appreciated his practicality. If a problem could be solved with the addition of a lead weight or an adjustment to the fuel mix, then that's what Mitchell would do, rather than return to the drawing board for lengthy analysis and re-tooling. Mitchell believed in achieving desired goals by the simplest means available.

On 7 September, in front of a huge home crowd crammed on to Southsea beach with a host of dignitaries, including the Prince of Wales and Lawrence of Arabia, watching from liners and yachts, the S6 saw off the competitors with ease. The combination of Mitchell's flair and Rolls-Royce's supercharged R-engine was a triumphant success, taking the 1929 trophy and reaching a top speed of 328.63 mph.

That moment of elation would have marked the end of the development of Supermarine's high-speed planes, but for the intervention of a showgirl turned multi-millionaire. By 1930, Britain was in the depths of the Great Depression. The Labour government under Ramsay MacDonald regarded the Schneider Trophy challenge as a frivolous expense that hard-pressed taxpayers shouldn't be asked to fund. An Air Ministry communique of January 1931 was quite clear on the issue:

> The Government has decided that in the present financial situation the expenditure is not justified.[15]

R. J. Mitchell was disappointed. He told the *Stoke Evening Sentinel* that British aircraft were the best in the world, but, 'if we drop our research work now and allow things to drift, in a year or two's time we may have lost that position.'[16]

If Mitchell was concerned by the decision, aviation enthusiast and ardent patriot, Lady Lucy Houston, was absolutely furious:

> I am utterly weary of the lie-down-and-kick-me atti-tude of the Socialist Government. To plead poverty as a reason for objecting to England entering a race against teams supplied by nations much less wealthy than our own is a very poor excuse.[17]

Lady Houston was not a woman to cross lightly.

Born Fanny Lucy Radmall, the ninth of ten children of a warehouseman and a seamstress, Lucy had the looks and spirit to escape her roots and thrive in the social whirl of late Victorian London. Young Lucy took to the stage as a dancer and at sixteen (or, as gossip suggested, considerably younger), she became the mistress of an heir to the Bass Brewing dynasty. He died in 1882 and, much to the fury of his family, left Lucy a bequest of £6,000 per year. It was a comfortable income for the time, but Lucy wasn't finished yet. She next married the son of a baronet, divorced him and married Baron Byron, a distant relative of the Romantic poet.

With her fortune and social status cemented, she enthusiastically threw herself into the cause of female suffrage, reportedly teaching parrots to screech 'Votes for Women' and paying Emmeline Pankurst's prison bail. On the death of Byron, she pursued the curmudgeonly shipping magnate Sir Robert Houston, married him and forced a re-write of his will. When he died in 1926, the extraordinary Fanny Radmall paused for just a moment, took a deep breath, and found herself sitting on a most agreeable fortune of £5.5 million. This was the kind of money – worth around £300 million today – that confers a certain degree of power.

As an ardent anti-Communist, Lady Houston was convinced that Labour Prime Minister Ramsay MacDonald was a Soviet agent and did everything she could to undermine his government. Their refusal to fund the Schneider Trophy challenge of 1931 caused the *Daily Mail* to label the government as 'socialist spoilsports'. Lady Houston spotted an irresistible opportunity to embarrass MacDonald. In return for the use of RAF pilots in the Schneider challenge, she would personally pay the government £100,000. She declared that: 'Every true Briton would rather sell his last shirt than admit that England could not afford to defend herself.'[18]

Recognising that they'd been outplayed, the government folded and threw its weight behind one more challenge. The Prime Minister swallowed his pride and offered his apparently unequivocal public support:

> I am sure that the result of the flight will be to demon-
> strate once again the magnificent courage and ability
> of our air force, for every member of which I have the

warmest personal regard, and to put beyond question the superiority of British engineering skill.[19]

Mitchell now had just nine months to prepare a fresh joust for the trophy. A brand new design was out of the question, but Rolls-Royce worked to squeeze more power out of their R engine while Mitchell turned the floats into giant radiators to dissipate the extra heat produced by the increased power and speed. Something like half of the aircraft's surface was now being used to cool the engine. Mitchell nicknamed his plane 'the flying radiator'.

Accidents and mechanical delays in the US, France and Italy plagued Supermarine's competitors. Italy entered the stunning Macchi MC72, but struggled to come up with an engine to compete with the Rolls-Royce R. Eventually their engineers resorted to the unorthodox technique of bolting together two Fiat V12 engines to turn two contra-rotating propellers. The resulting design was a beautiful death-trap, killing two of Italy's finest pilots. The Schneider Trophy had a dangerous reputation – a significant part of its attraction for the crowds who came to watch the races. By 1931, however, the speed of the planes paired with a nationalistic determination to win at any cost was proving a fatal combination. Early development work ahead of the race had already killed Italian, French and British pilots. But Battle of Britain pilots would one day have cause to thank these men who sacrificed themselves on the altar of speed.

By the eve of the 1931 competition, the attrition rate of planes and pilots ensured that the two Supermarine S6Bs and an S6A (the 1929 aircraft retro-fitted with the new engine) were the only aircraft ready and able to fly.

On 13 September 1931, watched by Lady Houston from her steam yacht, *Liberty*, the winning circuit was flown around the Solent by Flight Lieutenant Boothman in an S6B. With this third consecutive win for Mitchell and the team, the trophy was awarded permanently to Great Britain and the compelling but deadly Schneider series came to an end. Seventeen days later Flight Lieutenant George Stainforth broke the world speed record in the S6B, reaching 407.5 mph. In the space of just ten years the air speed record had been doubled.

R. J. Mitchell was relieved for a number of reasons. No more pilots would die in pursuit of Schneider's cup, Britain was triumphant and – most importantly for him – he had the dataset he needed to produce a new generation of aircraft. The BBC asked Mitchell to prepare a radio talk about his victory. For most of the British public it would be the first and only time that they would hear his voice. His son Gordon described how, conscious of his slight stammer, Reg practiced the speech, reading it aloud again and again to his wife, Florence. In his warm, West Midlands accent, accentuated by the occasional upper-class affected vowel, he gave listeners to the BBC National Programme a rather technical talk on the S6B, highlighting its advanced engineering and his own design philosophy:

> It is not good enough to follow conventional methods of design. It is essential to break new ground and to invent and evolve new methods and new ideas.

He described the S6B's innovative cooling system in great detail before reaching his dramatic conclusion:

For the present however it is generally considered that high speed development has served its purpose. It has accumulated an enormous amount of information that is being used to improve the breed of everyday aircraft.[20]

A new era of flight had been prised open and Mitchell and the Supermarine team had gained all the high-speed experience they would need to build the Spitfire. As he explained to a reporter from Southampton's *Southern Daily Echo*: 'Speeds which amaze us today will be the commonplace of tomorrow.'[21]

*

Europe was drifting inexorably toward war. Britain's fighter squadrons had to be equipped with planes that could reach the speeds that stunned the Schneider crowds. Mitchell and those brave pilots of the RAF High Speed Flight had made that possible. Mitchell and his loyal team of bright young designers were in the perfect position to build on their experience and seize the initiative for Britain.

When test pilot Jeffrey Quill joined Supermarine to work on the early prototype of the Spitfire, he quickly came to realise the importance of the Schneider experience:

Mitchell and the team which he had built up around him – which was incidentally a very young team – these chaps after the S4 and the S5 and the two S6s and all the world speed records, these fellows knew more about high-speed aeronautics than any other design team in the world.[22]

One intriguing footnote to Britain's Schneider success was revealed in 2020 by the Formula One racing engineer Calum Douglas in his book, *The Secret Horsepower Race.* Francis Rodwell Banks was a British expert in aviation fuel who worked for the Ethyl Export Corporation. His careful choice of the correct fuel mixture for the Rolls-Royce R-engine was instrumental in squeezing the very best performance from the S6B. These racing engines with a very short lifespan needed the optimum cocktail of fuel to achieve their maximum power output. For the S6B, Banks mixed 10% acetone with 30% benzole, 60% methanol and a pinch of tetra-ethyl lead to boost the octane rating. In a move that at first appears rather odd, the British government offered his assistance to the struggling Italian aero racing team. In Italy he learned a lot about the strengths and weaknesses of their aviation sector and made contacts that resulted in an invitation to give a series of technical talks in Germany.

In early 1933 he visited Berlin just as Hitler was planning the development of Germany's military aircraft industry. Banks offered his German counterparts in engine development the benefit of his advice on fuel tests and the best valve materials to use. He met the Reich Commissioner of Aviation, Hermann Göring, several times and even enjoyed two official dinners at his home.

Banks published an autobiography in 1978 that described some of his wartime work, but it was only with the release of confidential government papers that Calum Douglas was able to prove that Banks was travelling – and spying – on behalf of British Intelligence.[23]

The post-war assessment of Britain's slow response to German rearmament and her government's policy of appeasement toward Hitler can blind us to the fact that furious efforts were underway beneath the surface to keep British aviation ahead of the game. Perceptive politicians and influential individuals in the aeronautics industry were well aware of what was happening in Germany and the danger Hitler posed to peace. In the few, short years between the final award of the Schneider Trophy to Britain and the outbreak of war, they would help R. J. Mitchell develop his ideas into a brilliant fighting machine. The legitimate parents of the Spitfire weren't just engineers in a Southampton office block; they were spies, test pilots, champion motorcyclists and even a thirteen-year-old schoolgirl from north London.

9

A New Fighter Needed

As the Schneider Trophy was entering its final years, the design of fighter aircraft had barely moved on from the Sopwith Camels and Bristol F2s of the First World War. Bombers, however, had got bigger, faster and deadlier. The gap seemed impossible to bridge, a situation summed-up by three-term Prime Minister Stanley Baldwin in a speech to the House of Commons, in which he declared that 'the bomber will always get through'. The winner of a future war would be the nation best able to bomb enemy cities into submission. The ease with which Royal Air Force bombers had destroyed the forward bases of local forces in the Third Afghan War of 1919 and forced the surrender of rebel villages in Mesopotamia in 1923 provided ample evidence. Bombers could subdue an enemy with minimal risk to the aircraft and crew. The Air Minister, Lord Thomson, explained how he had dealt with one set of Mesopotamian 'recalcitrant chiefs':

> As they refused to come in, bombing was then authorised and took place over a period of two days. The surrender of many of the headmen of the offending tribes followed.[1]

Baldwin's view was widely shared within the British military establishment and bomber development was given priority. However, there was a small but influential group in the Air Ministry – among them the future leader of Fighter Command, Hugh Dowding – who had watched with interest as the Schneider Trophy spurred the development of small, fast aircraft. They believed that experience of the bomber's role in far corners of the Empire was of limited relevance to any air war fought in British skies. Fast, nimble interceptors had the potential to provide some measure of defence for Britain's airspace.

*

Dowding was a Scottish artillery officer who had learned to fly before the First World War and served with the Royal Flying Corps as a Flight Commander. As the inter-war Air Ministry's Head of Supply and Research, he had a pivotal role in setting out the specifications for new aircraft to be purchased by the Royal Air Force. Although he had no scientific training – and would later develop a passionate interest in spiritualism and fairies – Dowding was credited with an instinctive understanding of the latest in defence technology.

On 1 October 1931, less than a month after Mitchell had won the Schneider Trophy outright, the Air Ministry issued a formal invitation to manufacturers to submit plans for a fast fighter capable of a high rate of climb and high-speed performance at an altitude of 15,000 feet (4.5 km). It should be made of metal and armed with four 0.303 inch machine guns. The government preference was for a monoplane. Specification F7/30 was just the tonic that Supermarine and Britain's other aeroplane builders, mired in the depths of a global depression, needed. Eight companies entered with twelve different designs.

R. J. Mitchell was now very much top dog at Supermarine. The company had been sold to Vickers, an enormous engineering conglomerate, on the understanding that Mitchell came as part of the package. Mitchell was given a seat on the Vickers board and the unquestioning support of Vickers Aviation's fearsome chairman, Sir Robert McLean. By 1931 Mitchell had fifty draughtsmen and ten technicians working in his drawing office. He was no longer a niche designer working for a small-scale builder of flying boats – Mitchell had the resources of an industrial giant behind him and the creative freedom to come up with an imaginative response to the new fighter specification.

The Air Ministry was not, at least initially, disappointed. While Sidney Camm at Hawker offered a souped-up version of the existing Hawker Fury biplane, Mitchell showed them plans for a revolutionary gull-wing monoplane. This radical direction of travel appealed to the Ministry. A prototype was ordered, but the Type 224 encountered problems as soon as it left the drawing board.

The plane was designed around Rolls-Royce's new Goshawk engine. It was an innovative and powerful propulsion system, but it required elaborate evaporative cooling equipment. The cooling water could reach a temperature *above* boiling point by keeping it under pressure as it passed through the engine. Once it had left the engine the water would turn to steam, the steam would be cooled in a condenser and the resulting water circulated again through the engine. It was an ingenious system that appealed to the former railway engineer in Mitchell, but when fitted to the Type 224 the pressure proved unpredictable and the super-heated water had a tendency to turn to steam in the water pumps, causing them to seize up.

Mechanical problems could be solved, but, just as the design was being finalised, Mitchell was diagnosed with bowel cancer. After complex surgery and a short convalescence in Bournemouth, he returned to work, now reliant on a colostomy bag, but determined to keep his condition to himself.

Mitchell had always had a short temper. New staff members were warned to keep an eye on the back of his neck. If they had to enter his office while he was deep in concentration at the drawing board, they could gauge his mood from his neck. If it was turning red they should make a hasty exit. Good ideas would be rewarded and shared, bad ones might literally be ripped to shreds in front of the entire drawing office. In his largely respectful biography of his father, R. J.'s only child, Gordon, wrote of this unpredictability, the long hours he would work and his moments of rage.

His temper was now noticeably shorter and the rages more intense. Colleagues commented on the change, most completely ignorant of the cause. He was often in great pain and there was a strong chance that the cancer would return to kill him. If Mitchell had lost his passion for this fighter project, it would be understandable.

The Type 224, with its fixed undercarriage and ungainly bent wings, resembled a Steampunk Stuka. It was as ugly as the Spitfire would be beautiful, an aesthetic world away from Mitchell's elegant Schneider designs. In the air it lived down to expectations, struggling to reach 230 mph and failing to impress Supermarine test pilot Jeffrey Quill, who cruelly suggested that it was a plane that needed a plumber rather than a pilot:

It was disappointingly slow and had a poor rate of climb. The evaporative cooling was a dog's breakfast. It was just not a very good design. By hindsight the wing was too thick and big, the all-up weight too high and the drag too great.[2]

Frustrated by Supermarine's slow progress, the exasperated men from the Ministry had little choice but to order a more familiar shape – the Gloster Gladiator biplane – into production.

For both Sidney Camm and R. J. Mitchell, the failure to win the contract was a badly needed wake-up call. Hawker's Camm finally abandoned biplanes and Supermarine's Mitchell turned back to his Schneider-winning designs for inspiration. Both returned to the Air Ministry with fresh proposals. Both promised fighters that would be significantly faster and more manoeuvrable than the Gladiator.

On 14 October 1933, Hitler withdrew Germany from the League of Nations and the Geneva Disarmament Conference. Any lingering doubts about Germany's intention to rearm were banished. The need for a new generation of fighters to defend Britain was now urgent. Prototypes of both aircraft were ordered. R. J. Mitchell, still in recovery from cancer, would need every ounce of his creative genius and engineering expertise to turn his latest drawing-board ideas into something very special. He would also need help from some of the finest young aviation minds, at Supermarine and beyond.

*

One man at the Air Ministry was particularly concerned with the way in which British fighter development had been progressing. Fred Hill was a gunnery expert. In the First World War he worked at the Isle of Grain seaplane base in Kent, developing gunsights and trialling new weapons for the Navy and the Royal Flying Corps. By 1930 there were very few people in the world who knew more than Fred about the complex science of hitting a moving aircraft with a stream of bullets.

In 1931 Hill helped organise a first in British aviation – the Martlesham firing trials. An aircraft towed a target, which was attacked by a fighter armed with the standard twin Vickers machine guns. The trials quickly revealed the scale of the task facing the designers of the new generation of fighter aircraft. RAF aerial combat strategy of this period was to open fire from the maximum range of 1,000 yards or 900 metres, much as a naval cruiser or battleship might do. The trials showed no hits at all at this range. Hill became convinced that the only chance of a fighter taking down a fast bomber was to get close and fire as many bullets as possible as quickly as possible.

Fred Hill's superiors accepted his basic premise and specifications for new aircraft were altered to include four guns rather than two. With four modern guns firing at around twice the rate of First World War weaponry, that would surely be sufficient? Hill wasn't so sure. Fighters flying at 200 mph against bombers at the same speed might succeed, but the pace of engine development meant that the next generation of aircraft would be pushing 400 mph, giving fighter pilots just moments to make their attack count.

Hill needed to prove his point before the new fighters left the drawing boards and entered production. But how could he do it? He already had an important job, developing a new deflector gunsight for RAF fighters – any extra work on the aircraft's armament would have to be done in his own time. The maths involved was complex, with a mountain of data from the latest Martlesham trials to be sorted and crunched. His solution was simple, daring and almost certainly in breach of the Official Secrets Act. He took the figures home, along with the Ministry's latest computational device, and set his thirteen-year-old daughter to work.

Labelled as naughty at school, Hazel Hill struggled with words but handled numbers effortlessly. At the kitchen table in their terrace house in Highgate, north London, Fred and Hazel fed figures into the hand-cranked calculator. The innovative accumulator memory of the calculator made it possible to store data as they went along and Fred and Hazel gradually plotted out the relationship between aircraft speed, firing range and density of fire. The results were startling.

Hazel and Fred's calculations proved that a fighter would have to get close, very close, to have any impact. It would have to be manoeuvred into a position just 230 metres from the target. At that distance, chasing a fast, powerful bomber, the pilot would have the enemy in his sights for just two seconds. With four of the new Browning machine guns, each capable of just over 1,000 rounds per minute, that would put around 130 bullets in the bomber. Fred didn't think that was enough to bring down a bomber. Eight guns, carefully calibrated, would put 260 bullets into the enemy. That *should* do the job.[3]

In September 1933, Claude Hilton Keith took over as Assistant Director of Armament Research and Development at the Air Ministry. He was a Canadian veteran of RAF campaigns in Iraq and Palestine and had led his own fighter squadron. Keith immediately grasped the importance of Fred and Hazel's work, but persuading his superiors proved to be a tougher task. Hill was a quiet backroom number-cruncher, not the type to persuade fighting men of the First World War, so Keith enlisted two rising stars in the Ministry with solid combat experience. Air Commodore Arthur Tedder, the new Director of Training, had a particular interest in gunnery. He was recruited to Keith's team, alongside Squadron Leader Ralph Sorley of the Operational Requirements Branch. Sorley already acted as the bridge between the RAF and the manufacturers at Hawker and Supermarine, so it was vital to get him on board. Persuaded by the data, Sorley and Tedder presented Hill's figures at the Ministry offices in the summer of 1934. The meeting seemed to go well, but it would take a little longer for the eight-gun policy to percolate its way through the bureaucracy. The Air Ministry was still deeply split on where to spend its money – fighters or bombers, around the Empire or defending British shores? The question of four guns or eight must have seemed like a minor detail.

Fortunately for the development of the Spitfire, one senior figure in government had a clear vision and the cash to back it up. Neville Chamberlain, the Chancellor of the Exchequer in the predominantly Conservative National Government, is often painted as an arch-appeaser of Hitler's ambitions, but in the mid-1930s he was deeply concerned by the growing threat from the Luftwaffe. Chamberlain strongly believed that the first job of the Royal Air Force should be to protect the British Isles.

Money was funnelled into the Metropolitan Force, the RAF's front-line defences around London, and fighter development was now given the highest priority.

Since the failure of the Type 224, R. J. Mitchell and his young team had been busy developing their alternative proposal. The new plane would have to be smoother, less clumsy. The wings would be thinner, the wheels would retract and the cockpit would be fully enclosed. Most important would be a new engine. The Goshawk was a possibility, but in the Type 224 it had proven to be a complex beast with limited scope for improvement. Fortunately, Rolls-Royce had another engine in development. The company's chief engineer, Ernest Hives, had been the brains behind the R engines that had won the Schneider Trophy. With their short lifespans, those engines couldn't be the answer for a hardworking fighter, but Hives could offer the 27-litre PV12, a new engine with 1,000 horsepower potential. Add in a revolutionary new cooling system, devised by Frederick Meredith at the Royal Aircraft Establishment, and the foundations of something special were in place.

The new aircraft was dubbed the Type 300 and the Air Ministry was informed of its attributes. In January 1935, Supermarine received a new Air Ministry Specification F37/34. It requested the manufacture of a prototype of Mitchell's impressive-looking new aircraft. It must be fitted with the new Rolls-Royce PV12 engine and should be armed with *four* Browning 0.303 machine guns.

In April 1935, yet another Air Ministry Specification for a new-high speed fighter was issued, but this time the competition was opened up to any British aircraft manufacturer. At long last the influence of Fred and Hazel's figures could be traced. The first paragraph of F10/35 specified that the new aircraft should:

> Have a number of forward firing machine guns that can produce the maximum hitting power possible in the short space of time available for one attack. To attain this object it is proposed to mount as many guns as possible and it is considered that eight guns should be provided.[4]

That was precisely what Squadron Leader Sorley – along with Keith and Tedder – had been pressing for, but he could see no obvious reason why a completely fresh tendering process should be necessary. If the two fighter prototypes already in development could possibly be adapted to the new specifications, then any new tender would prove to be an expensive waste of time. On 26 April, Sorley visited the Supermarine works in Southampton. He raised the idea of an eight-gun Spitfire and found Mitchell surprisingly amenable. He reported back to the Ministry that Mitchell 'is naturally desirous of bringing the aircraft now building into line with this specification. He says he can include four additional guns without trouble or delay.'[5]

Mitchell had decided to equip his new prototype with an elliptical wing in place of the original straight design. Much thicker at the root where it joined the fuselage, the new wing shape was comfortably able to accommodate the four machine guns in each wing as well as a fully retractable undercarriage.

Hazel and Fred's long nights at the kitchen table had not been in vain. The new specification arrived just in time for Mitchell and Camm to adapt their plans. The Hurricanes and Spitfires that fought the Battle of Britain would be armed with eight machine guns, not four.

*

Squadron Leader Allan Scott joined the RAF in 1940 as an eighteen year old and flew Spitfires in the Siege of Malta. Shortly before his death in 2020, he told the BBC just how important Hazel and Fred's mathematics had been:

> The four would have been really difficult because it wouldn't have given us the firepower that we had. I'm awfully glad she was able to work that out for us. I never thought it would need that. It's just putting guns on and firing them – you never think that mathematically they had to be coded and work out the firepower.[6]

Hazel would have her own reasons to thank the Spitfire. At the outbreak of war she was studying to be a nurse. At her father's insistence she had stayed close to home, training at the Royal Free Hospital in Hampstead, but the hospital authorities decided to evacuate their trainee nurses 500 miles north to Marischal College in Aberdeen. It should have been a much safer spot than north London, but in late 1939 and early 1940, the Luftwaffe was engaged in small-scale prodding raids on Britain's east coast, testing out the country's defences. Hazel was walking home from lectures along Union Street, the city's main drag, when a bomber roared up behind her and began strafing the shoppers. She threw herself clumsily to the granite pavement, covered her ears and then looked up to see a very welcome sight: 'A nice young man in a Spitfire was chasing him away.'[7]

*

Hazel's role in the Spitfire story shouldn't give the impression that the plane's passage from Mitchell's drawing board to wartime legend was achieved purely by lucky breaks and clever schoolgirls. This was a big budget industrial project on a very significant scale. Success required the engineering might of Rolls-Royce to develop the PV12 engine – a masterpiece that would soon be renamed the Merlin. It also needed the accumulated expertise of Britain's top aeronautics researchers at the Royal Aircraft Establishment to resolve teething problems at its birth and complex operational challenges at the height of the Battle of Britain.

There were two crucial individuals at the Royal Aircraft Establishment – one certainly a heroine, the other, arguably, a villain – without whom the Spitfire could never have soared to such extraordinary heights. Our heroine enters the Spitfire story when it's already flying, fighting – and stalling – in the Kentish sky, but our villain provides a very welcome early boost to Mitchell's prototype.

10

The Spitfire Spy

Aero engineer Frederick Meredith is looking forward to a peaceful Christmas with his wife, Gwen, and their baby son and young daughter. In his work with the Royal Aircraft Establishment, he's made a vital contribution to the success of the Spitfire. Now he lives in a leafy corner of Cheltenham, developing his own auto-pilot system for the next generation of aircraft and missiles. Tall, thin and intense, he's a central figure in Britain's defence industries, described in a confidential government report as 'little short of a genius'. Frederick is going to need all of his intelligence, his charm, and a good deal of luck over the next couple of hours. He's been urgently summoned to the headquarters of Gloucestershire Constabulary and he's reasonably certain that this is a little more serious than a speeding ticket.

The Royal Aircraft Establishment's Farnborough home had been the centre of British aeronautics since the turn of the twentieth century. In 1904, the Army Balloon Factory moved from Aldershot to Farnborough Common in Hampshire, to gain the space needed to inflate the latest dirigibles, or airships. It was here too, on 16 October 1908, that the Wild West showman and pioneer aviator, Samuel Franklin Cody, made Britain's first powered aeroplane flight. The Balloon Factory morphed into the Army Aircraft Factory, and then the Royal Aircraft Factory, as it became obvious that fixed-wing aeroplanes were a more flexible and economical option for military use than balloons and airships.[1]

Naval power had been the top priority for Britain's military chiefs for centuries. The pre-First World War arms race with Germany had focussed almost exclusively on the Dreadnought battleships, while the new field of aviation research received only the most meagre of government crumbs. That changed overnight with the declaration of war in August 1914. The Farnborough facilities suddenly buzzed into life, turning prototype aircraft into production models that could be built by a new breed of manufacturers around the country. Staff numbers rose from 100 in 1910 to 4,000 in 1915, and two wind tunnels were built to improve the aerodynamics of the primitive fighting aircraft of the early war years.

Starting from such a low base of development, this first generation of professional aeronautic researchers had a fresh field of science and technology to explore. They calculated the advantages of all-metal fuselages, developed flap technology, air brakes and the first practical aeroplane compass. As the Western Front solidified into a stalemate of mud, barbed wire and trenches, the aeroplane took on a new importance – it became an artillery spotter, a bomber and, eventually, a dogfighter. The Royal Aircraft Factory team responded to the changing battlefield with bomb sights and super-charged engines. By the end of the war, Britain's aircraft were as fast, reliable and deadly as anything the Germans had, thanks in large part to the scientists and engineers of the Factory, which was renamed the Royal Aircraft Establishment in 1918. When the Armistice finally came, ex-RAE researchers spread far and wide, building the foundations of a new civilian industry. Hi-tech start-ups exploded across southern and central England, developing new aircraft and new engines.[2]

In the run-up to the Second World War, the Royal Aircraft Establishment cemented its vital research role for government and private industry, and the Spitfire gained immeasurably from RAE expertise. The most important early figure in this story was Frederick William Meredith, a young and idealistic mathematician from Dublin. In August 1935, his research in the RAE wind tunnels led to the publication of a hugely influential study, snappily titled, 'Note on the cooling of aircraft engines with special reference to ethylene glycol radiators enclosed in ducts'. The scientific paper was based on his observation of the ram-air effect, an aeronautical anomaly that would provide the Spitfire with a crucial boost.

High-performance engines need to be fitted with a system to regulate the temperature to prevent over-heating when they're working at their hardest. Meredith realised that the process of cooling could actually create a very helpful side-effect. Liquid-cooled engines, such as the Spitfire's Merlins and Griffons, operate in a fairly simple way. A cool liquid is passed through pipes in the engine, absorbing excess heat and keeping the engine itself at a safe operating temperature. The heated liquid is in turn cooled by outside air directed from the airstream over the pipes. The cool air becomes hot and is vented out as exhaust gases from the rear of the engine. Meredith believed that the expulsion of hot air was a significant waste – a loss of energy that could be useful in the air-craft's propulsion.

The secret of what became known as the Meredith Effect lies in the design of the air duct. This is the chamber on fast piston-engined fighter aircraft through which cool air enters and meets the radiator that's transporting liquid coolant heated by the engine. On the early models of the Spitfire, the air scoop for the ducted radiator was situated under the right-hand wing. The air scoop is essential to bring in sufficient cold air to the radiator, but, aerodynamically, it's an annoyance, disturbing the smooth lines of the Spitfire. Frederick Meredith worked out that a well-designed duct could compress the air. When the air passes through the radiator, it is heated and increases in volume. The hot, pressurised air then exits through the exhaust duct, which tapers to a small hole. The air accelerates backwards and provides forward thrust for the aircraft.

Meredith's work is most apparent on America's best single-engine fighter of the war, the P51 Mustang, with its distinctively enormous air scoop slung beneath the pilot. But even on the Spitfire, Meredith's ducted radiator was vital in compensating for the effect of drag and adding a crucial few miles per hour to its top speed. An early version of Meredith's radiator system was fitted to the prototype Spitfire K5054 that first flew on 5 March 1936, and the Meredith Effect would boost the performance of Allied fighters right through the war.

Meredith left the Royal Aircraft Establishment in 1938 to head up the Physics and Instruments Department at Smith and Sons, specialists in car and aircraft instruments. The list of his patents – for gyroscopes, servo-motors and radio direction and control – proves the breadth and speed of his important war work. Just like R. J. Mitchell, Meredith was no lone genius. His working technique was to thrash out technical problems by discussion and argument. He would argue for and against various hypotheses, encouraging his team of engineers to think for themselves. It was a working strategy that accelerated development just when Britain needed it most. He invented an early military drone – a remote-control aircraft, known as the Queen Bee, used to help RAF fighter pilots improve their high-speed shooting. Evacuated from north London to Cheltenham, Meredith and his twelve-strong research team created an autopilot system that became known as SEPI – Smiths Electronic Pilot No.1. It was revolutionary work that helped in the development of guided missiles as well as the autopilots on a new generation of heavy bombers and, eventually, the first British jet airliners.

On 23 December 1948, Meredith finished work for the day and drove from Smith's HQ out by Cheltenham Racecourse to his 5 p.m. appointment at the police station in Lansdown Road, close to the town centre. His tension increased at the turn of each mile. On arrival he was led into an office to meet a man who, from his dress and manner, was quite clearly *not* a member of the Gloucestershire Constabulary. The interrogation that followed is recorded in excruciating detail in MI5's extensive files on the case. Frederick was asked his nationality and how he felt about the British state. Where did he stand 'in relation to the State in an emergency'? It was obvious to Frederick where this line of questioning was leading. His left-wing sympathies were no secret. He exploded with pent-up fury. Why was he being asked these ridiculous questions?

What right did this man have to question him about his political beliefs?[3]

The MI5 officer remained utterly calm. He wrote later that Meredith was 'clearly a very frightened person, in spite of the belligerent stand which he took'. The officer showed Meredith photos of known and suspected Soviet agents. Meredith claimed never to have seen any of them before. By 6 p.m. Meredith seemed exhausted and asked if the interview might continue at his home just half a mile away. Astonishingly, the MI5 officer agreed. They drove in Meredith's car to the wealthy Park district of town. Meredith's wife, Gwen, 'a very pleasant young woman', according to the officer, agreed to take their daughter, Susan, and a visiting nephew out to the cinema but left their baby son, Johnnie, asleep in his cot.

As soon as the door closed behind Gwen, Frederick broke down. It was all true. He was a Soviet agent. From 1935 onwards he had become disillusioned with British foreign policy. He saw the political establishment splutter and roar with blind hatred at Russia while calmly allowing Hitler's Germany to re-arm. War was inevitable and it was vital that the Soviet Union should be ready to defend herself. When approached by the Russian secret service, Meredith said, he refused to actively 'spy', but agreed to supply any useful information that came his way.

Very little of this was news to MI5. The security services had been watching Frederick William Meredith for a long time. In the years after the Russian Revolution, the suspicion of Communist subversion had developed into a police and secret service obsession. Much of the industrial strife that accompanied the Great Depression was blamed – largely without evidence – on Russian inter-ference and Britain's small and ineffectual Communist Party was riddled with government infiltrators.

Meredith had been labelled as a Communist sympa-thiser for his organisational efforts in the General Strike of 1926 and support for hunger marchers in 1934. His pro-Sinn Fein views were noted, as were holidays and a work trip to the Soviet Union. His name was mentioned in a 1932 investigation into apparently extensive Commu-nist activity at the Royal Aircraft Establishment. A secret service insider at the RAE clearly took against Meredith, describing him to his handlers as 'an out and out Com-munist' and, rather more personally, 'a cunning swine'.

Despite his record, Meredith's extraordinary engi-neering talent kept him safe from purges of Communist agitators, leaving him in a position of some sensitivity when he was approached by representatives of the Russian secret service. According to his statement to MI5, he was persuaded by the journalist and Socialist activist Dorothy Woodman that it was his duty to offer something more than moral support to the cause. Woodman would later vehemently deny any role in Meredith's treachery.

Anxious to help the USSR in the approaching war with Nazi Germany, he agreed to travel home to Dublin. There, at the gates of Trinity College, he nervously scanned the crowds of students passing through the Georgian court-yard. He'd been told that his contact would be carrying a particular newspaper, sporting a pre-agreed flower in his button hole. The short, heavily-built man wearing rimless spectacles was not, as it turned out, hard to spot among the chattering, tousled and tweedy students. His newspaper and flower fitted the description. His clothes were French, but his accent was certainly Russian. This was Harry II, a spy master from the Soviet Embassy in Paris.

The first approach was a gentle one. Harry, also known as Andre, simply wanted Meredith's scientific advice. Any information that came his way that might be of use to Russia should be passed on. There was no suggestion at this stage that he should root out secrets or put himself at any risk. A one-off meeting was arranged between Harry, Meredith, a fellow RAE spy named Wilfred Vernon and a German middle-man, Ernst Weiss. From this point on Weiss would be Meredith's direct contact with the Russians. Their first rendezvous failed miserably on a dark and wet Surrey evening. Weiss and Meredith somehow missed each other as they waited with increasing impatience at opposite ends of the tiny village of Frimley Green. Soon, however, regular meetings established a smooth flow of secrets to Moscow.

Meredith claimed not to have been paid by the Soviet Union for his information, but his new life as a spy wasn't without its moments of glamour. Harry would occasionally drop by with gifts of French perfume for Gwen and when, in 1937, the young couple visited the USSR, Frederick asked Harry/Andre to pull a few strings. Sure enough, on arrival, Frederick and Gwen were plucked out of the lengthy passport queue and sped to their hotel. They were given VIP seats to watch Moscow's May Day Parade and treated to a beach holiday on the Black Sea.

The information Meredith passed on was intimately connected with his work. There's no evidence that he passed the secrets of the Spitfire to the Soviets, but they certainly received details of his research on stabilisation and aircraft remote control. Documents were passed to Weiss, who photographed them with his micro Leica, handed back the originals, and smuggled the tiny negatives to Paris. His Soviet handlers were reported to be disappointed with the quality of intelligence they were receiving. Meredith upped his game with a little light espionage, stealing a photographic slide of the schematic sketch of an automatic bomb sight. Harry was underwhelmed – the Soviets knew all about it already – and pressed Meredith to find direct military information on RAF squadron size and armament. Meredith claimed to have refused to supply this information.

A scribbled note on one of the many MI5 Meredith files implies a suspicion that some information gleaned pre-war by Soviet intelligence in Paris and London actually ended up in German hands. Meredith's refusal may well have been more important than he suspected at the time.

Despite his less than exemplary spycraft and no shortage of prying anti-Communist eyes at the Royal Aircraft Establishment, Meredith continued to pass information to the Soviets for at least three years. For two of those years he spied without suspicion, but an odd moment in a Surrey cafe brought him back on to the radar of the security services.

In early 1938, a bank clerk was taking tea in an Elstead cafe. He was intrigued by a sophisticated-looking young couple talking feverishly over a sheaf of foolscap documents. A page was turned and the clerk was convinced that he spotted the word 'Secret' stamped across one of the sheets. As a concerned patriot, he followed the couple out of the teashop, took note of their car registration and informed the local postmaster, who alerted the authorities. The car was registered to Frederick Meredith.

From now on that name was monitored carefully. A trip home to Dublin with his wife and some suspiciously heavy luggage provoked a thorough search of his belongings and a lengthy police interview at Holyhead port. His brother-in-law, a naval petty officer, reported Meredith's Communist leanings to his superiors, triggering a round of memos between different branches of the security services. Documents identify him as a clear security risk but his expertise continued to protect him. A May 1938 letter from the Air Ministry to Allan Harker, head of MI5's investigations and inquiries 'B' division, offered the opinion that any 'secret information in his possession emanates from his own brain'.

Ten years later, Meredith's treachery finally caught up with him. The Cold War had begun and the security services were urgently rooting out suspected Communist agents within the British defence industries. Ernst Weiss has been arrested and aggressively interrogated, surrendering his pre-war list of contacts. Of the Farnborough Soviet cell, Wilfred Vernon was now a Labour MP for Dulwich, and therefore to be handled with caution. Other former Leftist colleagues had moved on from their security-sensitive jobs. Meredith, however, was both fair game and of urgent interest. He was a senior researcher at Smith's in Cheltenham, a company about to be given contracts to make top-secret autopilot systems for guided missiles. It was vital to find out if Meredith remained a security risk. His phone was tapped, his mail intercepted and, with the written approval of the Prime Minister, he was called in for interrogation.

Frederick Meredith's MI5 interviewer at Christmas 1948 played a blinder. Meredith's past as a Soviet agent was personally confirmed and he appeared willing to do anything he could to keep himself out of prison.

Together they left his home and drove into the centre of Cheltenham for dinner. Both men seemed to have forgotten baby Johnnie asleep in his cot. Over dinner, Meredith expressed his irritation at the security service's pursuit of the small fry. Why did they 'chase mere typists at the Unity Theatre', rather than the Communists that Meredith believed to be much closer to the levers of power?

He openly discussed his politics and his true feelings about his country. He adored his wife and children and felt a strong loyalty to the company he worked for – he certainly didn't want to put them in an awkward position. He would be willing to resign his position if necessary, although he was not sure his wife would be very happy about that.

But what of that first question the interrogator put to him in Lansdown Police Station? Where did his loyalty lie? How would he react if Britain found itself at war with the Soviet Union? Such a situation, Meredith mused, would almost certainly be the fault of the United States. He would not betray Britain, but he could do nothing to harm the system he believed in. If he were asked to take on military work that would threaten the Soviet Union, he would refuse.

It was an extraordinary admission, but Meredith's candour seemed to both charm and reassure his interrogator. In his written report submitted on 30 December 1948, the MI5 agent offered his opinion that, despite his pre-war espionage activities, Meredith had no present connections to Soviet intelligence and too much to lose to re-engage with them. If the Cold War should turn hot, then he may have to be removed from sensitive work, but, in the meantime, his security risk was low.

On 9 May 1949, Meredith's MI5 file was stamped 'FILE CLOSED'. The man who gave the Spitfire a vital power boost and invented the forerunner of today's military drones may have been a Soviet spy, but he was also a brilliant engineer who played a key role in the defence of Britain at its moment of greatest weakness.[4]

11

Meanwhile in Germany

Meredith's treachery during the 1930s needs to be seen in the context of the developing global security situation. An international arms race was beginning. If Britain was poorly placed to defend herself, then the Soviet Union, with its long western border, was even more vulnerable. Mutual suspicion was high, but Britain and Russia shared a fear of the rapid rebirth of Germany's engineering and military might. On a state level, Britain was slow to react to the growing threat, but, as we've seen, individuals in government and the aviation industry were doing what they could to keep pace with a formidable and familiar foe. To appreciate just what the Spitfire's family of developers was up against, it's worth taking the time to understand the engineers who built those Messerschmitts and Heinkels that menaced the skies above Britain.

The roots of the aviation arms race lie in the Treaty of Versailles, which tried and spectacularly failed to cement peace after the Allied victory in the First World War. Germany's army and navy were to be strictly limited in size and firepower, there could be no German air force and there was a complete ban on the building or stock-piling of military aircraft. These restrictions were never broadly accepted by the German public, and clandestine rearmament took place under a variety of coalition governments through the 1920s. The German journalist and pacifist Carl von Ossietzky revealed the extent of this early illegal rearmament, publishing a 1929 exposé of M Section, a secret branch of the German army, which was training combat pilots in the Soviet Union. He was tried for the betrayal of military secrets and sentenced to eighteen months in prison. Von Ossietzky was awarded the 1935 Nobel Peace Prize for his efforts, but he couldn't travel to Oslo to receive his prize. He was already an inmate in one of the Nazis' first concentration camps.[1]

The Nazi Party had emerged from the 1932 general elections as the largest political force in Germany's parliament, the Reichstag. Adolf Hitler was appointed Chancellor on 30 January 1933 and, within three months, he had seized all the powers of a dictator. Opposition could now be wiped out and the extensive programme of rearmament that the Nazis had long promised could begin in earnest.

Throughout the democratic period of the Weimar Republic, a brilliant generation of young engineers had maintained Germany's strong position in aeronautical technology. When Hitler came to power, they were given a stark choice – co-operate with Nazi rearmament or surrender their positions and assets. Hugo Junkers, for example, was a pioneer of all-metal aircraft production. His Junkers W33 made the first east–west aeroplane crossing of the Atlantic and his inspired corrugated-metal Ju 52 would remain in production as an airliner and transport plane right up until the 1950s. Junkers refused to turn his company over to military production. Under threat of prosecution for treason, the Nazis pushed Junkers aside and took control of his thriving business. An example had been made of Junkers. For the sake of their careers, many others compromised and soon found themselves, whether sympathisers or not, locked into the increasingly formidable Nazi war machine.

While the Germans had continued to make strides in civil aviation during the 1920s and early 30s, the roots of the Spitfire, the Hurricane and the Messerschmitt 109 lay in the high-speed engine development promoted by the Schneider Trophy. The 1920 winner from Italy reached a speed of 107 mph. By the time Britain took the Trophy in 1931, the Supermarine S6B was reaching 407 mph.[2]

Germany had watched the Schneider Trophy from the sidelines, barred from high-speed aeronautics, but her engineers learned fast. Once freed from the constraints of the Treaty of Versailles, development of the Luftwaffe, and the new engines and aircraft to supply it, advanced swiftly.

Supreme German confidence in their industrial prowess was displayed in August 1937, with an invitation from the German Air Ministry to three of Rolls-Royce's top engineers. Would they be interested in a tour of the new engine research establishments?

The engineers – all deeply involved in the latest engine developments for Britain's fighter and bomber fleets – were met and entertained by top officials and given much more than a fleeting glimpse of German innovation:

We soon realised that Germany was thinking, planning and acting on a very much larger scale than we were . . . We saw and were allowed to examine all their engines and machines up to the latest production

. . . if the present rate of progress continues, which as far as we can see must be maintained because of the organisation behind it, we are definitely of the opinion that unless we ourselves think and organise on a much larger scale we must inevitably fall behind.[3]

Despite all the restrictions imposed on inter-war German industry, they had clearly caught up:

The workmanship is exceedingly good, and the external finish is very clean, particularly the Daimler Benz.

The team from Rolls-Royce had just seen the engines that would soon be engaged in desperate dogfights with their own Merlins.

There were more visits to Germany in October 1937, when an aeronautical convention hosted some of the top aircraft and engine designers from Britain and the United States. As at so many trade conventions, before and since, the drink flowed and inhibitions were lowered. In a Munich beer hall, the head of engine development for the German government, Helmut Sachse, admitted his distaste for Hitler and described how fear of the Führer's moods and caprices was hindering development. Aero engineers were too afraid of the consequences to admit to even the smallest failure. Sachse was soon removed from his position.[4]

Returning home from Munich, the British government's Director of Engine Development, George Bulman, came to a sobering conclusion:

In the quality of aero engine design Germany is on the point of equalling our best efforts and having achieved that result, starting from zero in 1933, she bids fair to surpass us completely very soon.[5]

But why did the Germans invite such perceptive representatives of a likely future enemy to see their advanced research and development facilities? They surely realised that anything the Rolls-Royce team saw or heard would be passed to British Intelligence? One interesting theory crops up in Calum Douglas's 2020 book, *The Secret Horsepower Race.* He uncovered a secret post-war memo sent to the Director of the CIA from an American visitor to the German factories and design workshops: 'There was wide speculation at the time as to why the Germans were showing their wares so obviously.' The memo's author rejected the idea that the Germans were simply puffing out their chests and showing off their superiority: 'These disclosures now appear to have been an "Operation Red Herring", to cover the really important research and development that was going on in the field of jet engines and guided missiles.'[6]

Britain too was engaged in the advanced technological developments that the Germans kept off-stage, but government support for Frank Whittle's revolutionary turbo jet engine was inconsistent and, at best, half hearted. With the likelihood of war increasing fast, Whitehall opinion favoured improvements in existing technology. In terms of fighter planes, that meant the Hurricane and the Spitfire. But if the Air Ministry thought they'd opted for the safer option, they would soon be disappointed. The Spitfire was far from ready for combat.

12

Data and Danger

In late 1935, Mitchell's Type 300 – the Spitfire prototype – was taking shape in the sheds at Woolston. The day was approaching when it would take to the skies for the first time. It would be a moment of relief, but no more than that. The toughest work would still lie ahead: a painstaking period of collaboration between Mitchell and a Supermarine test pilot to turn a plane with potential into a world-beating fighter. It was work that required a pilot with an exceptional range of skills.

The job of a test pilot in the 1930s and '40s was vitally important and incredibly dangerous. Few possessed the combination of qualities Jeffrey Quill identified for the BBC in 1976:

> I would say that the man who is without fear or claims to be without fear is a fool, not to mention a menace. And certainly there would be no place for such a man in test flying. What you want is a man who is subject to all the normal human reactions of this sort but is thoroughly well able to analyse them and control them and identify them.[1]

With no flight simulators or computers to crunch the data, the first moment that a designer could be sure that their new plane would fly at all was when a test pilot fired up the engine and took it for a spin.

As well as being a fine aviator, the test pilot had to be a meticulous data-gatherer:

> We used to have certain elementary recording apparatus. We had no air to ground telemetry or anything of that sort. We used to get a lot of built-in instrumentation, all sorts of special stuff festooned around the cockpit but you had to read it yourself and write it down.[2]

Jeffrey Quill described how he would have a clipboard tied to his thigh with a built-in stopwatch. While concentrating on the control of an unfamiliar aircraft, he would have to transcribe numbers from the dials around him, all the while using his flying experience to form an instinctive judgement on the flight handling characteristics that the instruments couldn't measure.

Quill was the second man to fly a Spitfire, the man who flew and tested every significant production variant of the aircraft. Nobody knew the Spitfire like Jeffrey Quill.

As a boy growing up in Littlehampton in West Sussex, he was excited when Royal Flying Corps pilots based at the nearby Ford airfield dropped by for lunch or drinks with his parents. The glamorous uniform and tales of derring-do hooked Jeffrey instantly. Before leaving his public school, he had applied to the Royal Air Force, joining at the age of eighteen. Most young men of his class would have applied for officer training at the RAF College at Cranwell, but Quill's father had recently died. To help support his mother, he wanted to earn a wage right from the start of his career, so he accepted a Short

Spitfire final assembly at Eastleigh, Southampton.

An air-raid shelter at Supermarine, which has been destroyed in the September 1940 raids.

Spitfire inspector
Dorothy Handel
of the Aeronautical
Inspection Directorate.

The Spitfire Sweethearts – Florrie and Daisy Snelling – riveting sisters
who married on the same day.

Winston Churchill discusses riveting techniques with a worker at the Castle Bromwich Spitfire factory.

R. J. Mitchell, the designer of the Spitfire.

Hazel and Fred Hill, the mathematicians who gave the Spitfire eight guns.

Spitfire test pilot, Jeffrey Quill.

The Experimental Hangar Netball Team from the Supermarine design hub at Hursley Park.

Spitfire trouble-shooter
and champion
motorcycle racer,
Beatrice Shilling.

Proud parents,
Forrest and Dinah,
show off little Michael,
the first baby born to a
British WAAF officer
and an American pilot.

Three Americans
who flew Spitfires
for the RAF in the
Battle of Britain
(*l to r*) Eugene Tobin,
Vernon 'Shorty'
Keough and
Andrew Mamedoff.

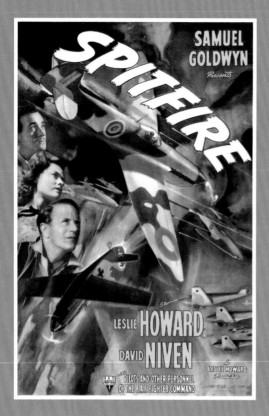

Above. Battle of Britain pilots,
Geoffrey Wellum (*right*) and
Brian Kingcome.

Left. Film poster for *The First of
the Few*, released as *Spitfire*
in the United States

Squadron Leader Mark Sugden
of the Battle of Britain
Memorial Flight.

Winnie Crossley

Joan Hughes

WOMEN OF THE ATA

Founder of the ATA women's section, Pauline Gower (*left in both pics*) with her pre-war business partner, Dorothy Spicer.

ATA pilots at RAF Brize Norton. From left to right: Mary Guthrie, Veronica Volkerz, Monique Agazarian, Rita Baines and Joy Gough.

Jack 'Farmer' Lawson with David Lloyd (*right*) and George Unwin (*left*) of 19 Squadron at RAF Fowlmere, September 1940.

Service Commission. He was swiftly spotted as an exceptional flier and was fast-tracked into 17 Squadron, a crack fighter unit flying the latest Bristol Bulldogs. He flew aerobatic displays for the public and seized any opportunity he could get to fly new aircraft being trialled by the Aeroplane and Armament Experimental Establishment.

Quill's next career move was into the RAF's specialist meteorological flight. Twice a day he would fly up to nearly 30,000 feet (9 km), where he would record temperature, humidity and cloud formations, landing to call the readings in to the Met Office in London. If low cloud prevented a return to his Duxford base, Quill would be forced to land his Armstrong Whitworth Siskin biplane in a field. His ambition on those occasions was always to find a flat and crop-free field, conveniently close to a telephone box. He would dial Holborn 3434, read out his results, and then sit back and wait for the clouds to pass. High altitudes, extreme weather and a daily dose of improvisation – this was ideal training for a test pilot.

As the autumn of 1935 turned to winter, Vickers Aviation's Chief Test Pilot, Joseph 'Mutt' Summers, was struggling with his workload. Hitler's aggressive intentions were now apparent to many in Britain's National Government. The Chancellor of the Exchequer, Neville Chamberlain, had released funds for air defence and there was a rush of orders for new aircraft from Vickers and its Supermarine subsidiary. The new Wellington bomber was taking up much of Mutt's time and the Vickers Vildebeest and Wellesley both required air testing. Summers urgently needed some help – ideally an experienced fighter pilot with skills in data collection. Jeffrey Quill fitted the bill perfectly and, in the dog-eat-dog world of 1930s arms manufacture, he had another appealing quality – Quill was terribly well connected.[3]

As a pupil of Lancing College, West Sussex, he had been educated alongside future politicians and military top brass, and his stint at the Met Flight based near Cambridge had turned into something of a social whirl. When not dancing with debutantes in London nightclubs, Jeffrey was a weekend guest at Port Lympne, an extraordinary Kentish country house said to combine Hollywood excess with faux-Tuscan elegance. A relatively modest home built in the Cape Dutch style, it had been remodelled after the First World War by the most fashionable artists and designers of the day. In the words of the architect Philip Tilden, Port Lympne was:

> the epitome of all things conducive to luxurious relaxation after the strenuousness of war. It was to be a challenge to the world, telling people that a new culture had risen up from the sick-bed of the old, with new aspirations, eyes upon a new aspect, mind turned to a new burst of imagination.[4]

This show home for a new culture of the imagination was the rural retreat of Phillip Sassoon. The Under Secretary of State for Air was a relentless networker, a man whose passion for flying was rivalled only by his delight in the company of the handsome young pilots themselves.

Weekends involved extravagant dinners, rigorous bouts of sea-bathing and road trips in Sassoon's fleet of Rolls-Royces, decorated with personalised hood ornaments. The familiar Spirit of Ecstasy was replaced on the bonnet with a striking cobra. At Lympne, Quill met great artistic figures of the day, such as Osbert Sitwell and Noel

Coward, as well as politicians and senior civil servants attracted by the glamorous company.

Jeffrey Quill joined Vickers Aviation in January 1936 to work under 'Mutt' Summers. Big and bold, Mutt had carved out a unique role in British aviation, styling himself as not just a pilot but an integral part of the company's design and development team. While most test pilots were employed to check that each new production aircraft was air-worthy, Mutt was a respected and influential voice in guiding the company's future projects. As Mutt's assistant, Quill found himself in meetings with the Vickers Aviation MD, Sir Robert McLean, Supermarine Chief Designer R. J. Mitchell and other senior members of the design team. Handsome, charming and sharp-witted, Quill soon cemented his own position as a core member of the team that would perfect the Spitfire. If Mutt belonged to the first generation of test pilots – rather disorganised, with a devil-may-care attitude to life and death – Quill was a foretaste of the technological future. He was meticulous in his methods and eager to understand the science and engineering behind each and every aircraft he flew.

It was Mutt who took Spitfire prototype K5054 for its maiden flight on 5 March 1936, a gentle eight-minute cruise with the undercarriage kept down. Quill reported that it was 'fairly obvious that Mutt Summers was very pleased with it'. There would be many more flying hours required before it was fully tuned and ready for delivery and combat training, but Mutt's skills were in demand elsewhere in the sprawling Vickers industrial empire. For the rest of the Spitfire's long life, it would be Jeffrey Quill who would test, appraise, analyse and test again as it developed through fifty-two different variants.[5]

Quill took his first Spitfire flight on 26 March 1936 at Eastleigh aerodrome near Southampton. His parachute was fastened and the cockpit closed. Everything, as he told the BBC in 1976, was already familiar:

> One had been for the previous three months working and looking forward to this day. I was absolutely familiar with the cockpit – I'd sat in it for hours, I could have touched anything blindfold and I had a very good briefing from Mutt who had already flown it six times by the time I flew it.

Once airborne, the easy sense of familiarity slipped away, replaced by something else entirely:

> It was something very different to anything one had ever flown before and I remember having a tremendous feeling of elation.[6]

This, though, was not the moment to let his emotions run free. Quill had a job to do:

> The aircraft began to slip along as if on skates with the speed mounting up steadily and an immediate impression of effortless performance was accentuated by the low revs of the propeller at that low altitude. The aeroplane just seemed to chunter along at an outstandingly higher cruising speed than I had ever experienced before, with the engine turning over very easily and in this respect it was somewhat reminiscent of my old Bentley cruising in top gear.[7]

Quill's apparently cool, calm and professional approach to his first Spitfire flight was, it seems, completely genuine. In 1970 he looked back on his younger, confident self with a wry smile:

> I was very young when I first flew the Spitfire, I was only 23 years old and I don't think I felt it half as much as I should have done. When I look back on it now and realise the issues that were at stake with that little prototype aeroplane, the only one we had, and for two years we flew. When I think of some of the things I did to it, it wakes me up in the middle of the night sweating.[8]

Supermarine's Chief Designer, R. J. Mitchell, must have felt even more anxious than Quill. As the world's only Spitfire was being tested to its limits, regular bouts of pain wracked his body. He knew that his cancer had returned and the prospects for recovery this time around were very slim indeed. Time to perfect the Spitfire was running short. As Quill took K5054 to the air again and again – recording speeds, cooling figures and the impact of different radiator settings – R. J. Mitchell was there watching and taking notes.

> I used to look down as I was on the circuit and I would nearly always see old Mitchell's yellow Rolls-Royce either driving in through the gate or parked on the tarmac. As soon as I taxied in he was always there. If we were going to fly it again he always used to invite me to sit in his car and talk to him.[9]

Ever since his days drinking and chatting long into the night with the High Speed Flight Unit, Mitchell had regarded the pilot's perspective as an essential part of aircraft development:

> He used to question me a lot about the aircraft. I was very young and not a trained engineer and slightly overawed by all these highly technical chaps at Supermarine. Old Mitchell was aware that I had a bit of an inferiority complex because I had no engineering degree or anything like that.[10]

The two very different men – one young and eager to express his opinions, the other urgent and anxious to pass on his knowledge – learned a lot from each other, and the Spitfire would continue to benefit from their short, intense relationship long after Mitchell had died:

> He made me understand the way engineers' minds work and I learned how to talk to them because after all a test pilot is a sort of bridge between the engineer on the ground and the aeroplane in the air.[11]

The figures that Quill was recording in these early flights weren't altogether encouraging. The aircraft would soon have to pass official trials set by the Air Ministry at the Aeroplane and Armament Experimental Establishment at Martlesham Heath in Suffolk. These trials would decide the Spitfire's fate. Neither Mutt Summers nor Jeffrey Quill had taken K5054 over 335 mph. That put it on a par with the Hawker Hurricane, an aircraft that was simpler to manufacture and maintain. If the innovative Supermarine plane was no faster than the Hurricane, then Ministry interest would quickly fade away.

A new propeller design cut the drag and Quill managed to squeeze an extra 13 mph out of the prototype.

That might be just enough, but Quill had other worries. There were the ailerons – the crucial wing flaps – which became heavy to control at high speed. The overall balance of the aircraft also worried him: 'The prototype was always teetering on the edge of instability which is a dangerous condition for a very fast aeroplane.'[12]

Neither the flaps nor the stability were likely to concern the Ministry. They wanted an aircraft to intercept high-flying, fast bombers, not a dogfighter. They needed something that current RAF pilots could fly without extensive extra training. At Martlesham Heath, RAF Flight Lieutenant Humphrey Edwardes-Jones took the controls of K5054. After a very satisfactory twenty-minute flight, he landed, called the Air Ministry and offered his positive impressions. The plane was fast but certainly manageable. That was good enough for the increasingly desperate British government. Just eight days later, on 3 June 1936, a contract was issued for the construction of 310 Spitfires.

It was a great day for Mitchell and his team, but celebrations at Supermarine were brief. The company's workers were used to the slow and meticulous construction of small batches of hand-built seaplanes. Mass producing a cutting-edge fighter, operating at a speed and altitude far beyond most existing aircraft, was a very different challenge. Add in the need to co-ordinate work from numerous sub-contractors and a chronic shortage of government-supplied equipment, such as standard guns and instruments, and the whole project appeared to grind to a halt.

Two long and frustrating years of intensive tooling, testing and tweaking followed. R. J. Mitchell would not live to see out those years. For months he did his best to ignore the inevitable, working long hours on the Spitfire and a new long-range bomber project. Eventually, he was forced to accept that his body could no longer cope. He retreated to his bed and wrote a heart-rending sequence of letters to his seventeen-year-old son, Gordon. A final desperate trip to a cancer specialist in Vienna did nothing to reverse his condition. After five weeks of intensive treatment, he returned on a private plane back to Eastleigh aerodrome, where the Spitfire prototype was based. He died at noon on 11 June 1937.

Jeffrey Quill flew the first production model of the Spitfire Mark I on 14 May 1938. The patience of the Air Ministry had been pushed far beyond breaking point. This beautiful aircraft had proven to be an awkward and expensive beast to build, and when Quill took to the controls he found that some of the original flaws were still there, some, in fact, had actually got worse.

Hundreds of the Spitfire's German rival, the Messerschmitt Bf 109, had been produced by this time. Although German government orders had been issued at around the same time as those for the Spitfire, the 109 had already been tested in combat in the Spanish Civil War and a modified racing model had broken the world speed record for landplanes. As the first 310 Spitfire Mark Is made their way through Supermarine's Woolston plant and war crept closer, it became obvious that Mitchell's fighter would have to adapt without him. An evolving threat would have to be matched. Quill would prove to be a vital figure in perfecting the Spitfire's handling, but others were needed to develop the engine and perfect that often forgotten factor in fighter planes, the fuel.

*

The standard aircraft fuel at the outbreak of war was 87 Octane, lower than today's standard for petrol cars. Raising the octane rating of a fuel doesn't in itself increase the engine's power, but it enables cylinder pressures to be increased without harmful detonation occurring. So, put simply, with the correct engineering in place, higher octane fuels will make Spitfires or Messerschmitt 109s fly faster.

Francis Rodwell Banks, technical manager of the Ethyl Export Corporation, gave a lecture in 1937 to the Institution of Petroleum Technologists and the Royal Aeronautical Society that set out the clear advantages of higher octane fuels for high-performance military aircraft. This fuel was complex and expensive to produce and, before the outbreak of war, there were no refineries in Britain with the capacity to make it. Despite this, the government's Oil Policy Committee decided in December 1939 to take a calculated risk. Spitfires would be fuelled only with 100 Octane petrol.

American oil companies had recently refined a British invention that promised a much faster way to produce higher octane fuels. Orders were made and stockpiles of the precious fuel from Esso and Shell plants in Texas, Louisiana, Curaçao and Aruba were built up just in time for the outbreak of the Battle of Britain. The US supplies gave vital breathing space, allowing Britain to build its own synthetic fuel plants that would help satisfy the enormous demand that continued to build toward D-Day and beyond.

From the outbreak of war, Germany was totally reliant for high-octane fuel on synthetic supplies derived from coal. Their grasp of fuel technology was arguably ahead of Britain's, but their domestic refineries simply couldn't produce enough of the high-octane petrol to keep the enormous German war machine moving. In the Battle of Britain, Messerschmitt 109s would fly on 87 Octane fuel. Britain's 100 Octane fuel gave the Merlin III engines a 30% power increase, from 1,000 to 1,310 horsepower. Compared with the 1000 hp of the Messerschmitt's Daimler Benz DB61 engine, it provided the Spitfires of the Battle of Britain with a small but significant power advantage.[13]

British engineers had been given a tantalising pre-war glimpse into the advances that German engine-makers had made in just a few years. They knew that they were pioneering techniques still untried on production aircraft in the UK, but technical detail was hard to come by. What they really needed was a German engine. On 28 October 1939, the first German aircraft to be shot down over Britain crashed into the Lammermuir Hills, to the south-east of Edinburgh. The pilot and navigator survived. More importantly for the British war effort, the Heinkel 111 was largely intact. Two men from Rolls-Royce – William Gill and Dick Royce – realised that this was the perfect opportunity to study a state-of-the-art German engine. It was vital that they reach the lonely spot before the RAF investigators. Once the site had been sealed, a thicket of ministerial red tape would have to be negotiated to gain access. They hired a truck with a crane and raced, as best they could, from Derby to the Scottish border.

Despite crowds of souvenir hunters, the Rolls-Royce team managed to remove one of the Jumo 211 engines from the Heinkel bomber. The engine was beautifully made, exploiting a fuel-injection system that markedly improved the engine's fuel economy. Fuel injection was an arena of engine development that Britain had neglected in the 1930s. That, studies of the German engine concluded, had been a serious error.[14]

It was too late now for significant changes in engine design. All the Rolls-Royce engineers could do was gaze with admiration at their enemy's achievements – replicating or surpassing them would have to wait for another day. In the meantime, the Merlin would just have to be tweaked and prodded to produce more power. The opening weeks of the Battle of Britain provided Supermarine and Rolls-Royce with a sudden rush of anecdotal evidence about the performance of their machines, but it was hard to separate facts from impressions. Were problems caused by the hardware or the pilot? What they really needed was reliable data. They needed real combat reports from real pilots with a skill for analysis rather than barrack-room anecdotes. Jeffrey Quill would give them just what they needed.

As Hitler's *Blitzkrieg* crushed Poland and rolled westward toward France, Quill had become convinced that he should do his bit on the front line. As a fully trained RAF pilot, he felt that he should be at the controls of a Spitfire, hunting down Messerschmitts, not tinkering with ailerons in the Hampshire countryside. His bosses at Supermarine strongly disagreed, so he went straight to the top and called Keith Park, head of Fighter Command's 11 Group, preparing to defend London. His pitch was strong, if a little mendacious: 'I concocted a story saying that I thought I would be able to do my job as a test pilot much better if I had some first-hand experience.'[15]

In truth, Park was never likely to turn down an experienced fighter pilot. On 5 August, Quill squeezed back into his RAF uniform and reported for duty to 65 Squadron at Hornchurch in Essex.

How sincere he initially was about returning to Supermarine with valuable combat insight is certainly questionable – like many pilots, Quill was anxious to get into action – but it turned out to be one of the most important postings in the Spitfire's history. From the first moment he faced the enemy, he saw the Spitfire from a fresh perspective. There was now a direct line of communication between the front line and the design team in Southampton.

Quill was given just a few days of combat training before flying with his new 65 Squadron colleagues to a dispersal aerodrome at Manston in Kent, geographically one of the closest air bases to Occupied Europe. It was so close to the Germans that the Manston Spitfires were forced to fly inland to gain height before turning back out to sea. If Jeffrey Quill really wanted to know how the Spitfire would perform in the heat of battle, then August 1940 at Manston was the right place to be.

On 12 August, 65 Squadron was forming up for take-off when there was a blast of cannon fire immediately above Quill's head. In his mirror he could see the hangar roofs behind him shudder and lift into the air. Throttles were slammed open and Quill's Spitfire accelerated:

> As I became airborne I glanced in the mirror and saw nothing but bomb-bursts and showers of earth and smoke immediately behind me. I thought I must be the only member of the squadron to have got away.[16]

Forty-eight bombs landed on the airfield, according to Quill's account, but miraculously eleven of the twelve Spitfires made it safely into the air. There was no chance to exact revenge, though, as the enemy was already roaring back across the Channel. Manston had been the victim of a specialist Luftwaffe dive-bombing unit, Erpobungsgruppe – or Test Group – 210.

Germany's Stuka dive bombers had been effective in destroying airfields and air defence systems in Poland and France, but it had quickly become apparent that over Britain they were easy meat for the faster Hurricanes and Spitfires. Erpobungsgruppe 210 was the Luftwaffe's solution. Led by a charismatic Swiss-German veteran of the Spanish Civil War, Walter Rubensdörffer, 210 was formed of bomb-carrying Messerschmitt 109s and 110s. They were tasked with lightning-fast hit-and-run attacks on coastal airfields and radar stations. Flying close to the wave tops, the E-210 aircraft could appear suddenly over a cliff edge and create havoc, escaping at speed before any surviving British fighters could take to the air.[17]

The laser focus of the Luftwaffe on Fighter Command's airfields, aircraft and communication systems had precisely the intended effect. In the two weeks from 24 August to 6 September, it's estimated that 295 British fighters were destroyed and 171 damaged. The Luftwaffe was losing fewer Messerschmitt 109s and had the production capability to replace them fast. It became absolutely critical that the surviving Spitfires operate to their optimum level.[18]

Once airborne, dodging and weaving with German fighters, the Spitfire's true combat performance revealed itself. Quill was no longer dealing with theoretical stresses and limits; he was pushing the Spitfire as hard as he could every day. He felt he had the edge over the Messerschmitt 109 in terms of rate of climb and a tighter turning circle, but his calls and letters to Joe Smith, R. J. Mitchell's successor at Supermarine, reported a host of problems.

In the high-speed dogfights that defined the Battle of Britain, it was imperative that you see the enemy before they saw you. In a Spitfire Mark I, your field of vision was significantly impaired. The side panels of the cockpit windscreen were made of curved Perspex, distorting the view, particularly over long distances. There were also problems from the engine cooling system, with coolant boiling over at low speeds, ejecting steam that could freeze on the windscreen. Rapid dives threatened stalls and caused more condensation:

> When you had been flying at high altitude for some time the inside of the windscreen got very cold. As you descended into warm, moist air the moisture condensed on the inside of the screen and promptly froze up . . . You couldn't damn well see.[19]

All this detail was fed back to Supermarine. Quill's reports also suggested the fitting of a screenwash system, to remove oil deposits, and a bullet counter, so that pilots could be confident that they had enough ammunition to press home an attack.

Most urgent of all, though, was the old problem Quill had spotted in the Spitfire's earliest days: 'The aileron control was very, very bad in the early Spitfires – much too heavy.'[20]

Again and again Quill and his comrades found themselves in high-speed manoeuvres, struggling to move the flaps and control the aircraft.

On 24 August 1940, with less than a month of front-line action under his belt, Jeffrey Quill was instructed to return to Supermarine. The Mk III was ready for testing and he was the only man for the job. Quill was deeply frustrated by the order, but in those few intense weeks he'd built up an invaluable bank of knowledge about the Spitfire's performance and a set of strong opinions on the best ways to improve it.

Returning to Hampshire, he was pleased to find the Supermarine engineers hard at work trying to resolve the problem of those heavy ailerons. He was set the immediate task of flying aircraft with a variety of different settings and fittings to the flaps. For weeks, nothing seemed to improve the situation, with many of the bright ideas actually reducing the aircraft's performance. On 7 November, Quill took off in a Spitfire fitted with metal-skinned ailerons with thin trailing edges.

Despite its promotion as a modern, all-metal aircraft, the Mk I Spitfire had so far clung on to a few features of heritage aviation. The ailerons and rudder were still constructed in the old traditional way, as developed by the Wright brothers at the turn of the century. Fabric was stretched over a frame and covered in 'dope', a chemical lacquer that stiffens the fabric and makes it waterproof. The switch to metal ailerons made an immediate and dramatic difference – the Spitfire was now a dream to manoeuvre at high speed. Production began on new ailerons immediately and, as soon as the word got out, combat pilots were clamouring to have their fabric flaps replaced.

While Quill and his colleagues struggled to correct mistakes in the original design of the Spitfire, the engineers at Supermarine and Rolls-Royce were working toward fundamental improvements in the next generation of Spitfires. German engineers had learnt lessons from the battles over Poland and France and they were quick to respond to the weaknesses highlighted over Kent, Sussex and London. Their British counterparts had to do the same.

Rolls-Royce already had a better engine ready to fly. The Merlin XX was fitted with a two-speed gearbox in its supercharger. This significantly increased its performance at high altitude. Unfortunately, the XX was noticeably longer than the Merlin III fitted to the Spitfires fighting over southern England. It wouldn't fit into the current Spitfire Mk I or the improved Mk III that Supermarine was preparing for service. A short-term bodge job was contrived, with Rolls-Royce producing the Merlin 45 engine that offered some of the XX's improved performance at altitude but in a package size that would fit the Mk I chassis. This mongrel aircraft was labelled as the Mk V. It put the Spitfire back on par with the latest variant of the Messerschmitt 109, but the high-speed arms race was getting even faster and tougher for the British.

Some of the Messerschmitt 109s flying at high altitude over Britain seemed to be able to give themselves a sudden and dramatic boost in speed. Crashed aircraft revealed a mysterious extra gas tank, and captured airmen made a reference to the new 'Ha Ha Process'. British Intelligence didn't quite live up to its billing on this occasion. Despite the obvious clues, they failed to realise that the Germans were injecting nitrous oxide, or laughing gas, into the Messerschmitt 109's supercharger. British engineers wasted many man-hours trying to recreate the effect in their own fighters with a much more dangerous oxygen boost.

Fortunately, one engineer was more perceptive and more focussed. She knew what really mattered – not dangerous gadgets or colourful war-winning wheezes – but a solution to the Spitfire's most persistent niggle, a problem that was allowing countless Luftwaffe pilots to escape with their lives.

13

Miss Shilling's Orifice

For a Battle of Britain pilot there was nothing more frustrating than the loss of a certain kill. Two of the top Battle of Britain aces – Douglas Bader and Robert Stanford-Tuck, with the help of some Airfix models – explained the problem rather brilliantly for a BBC film in 1976.

In their scenario, the Spitfire pilot has done the hard work exploiting his tighter turning circle to get on the German's tail. The enemy is perfectly framed in his cross-hairs. His thumb presses the firing button, but the Messerschmitt makes a well-practised manoeuvre:

> Stanford-Tuck: . . . as the German my immediate evasive action would be to just stick my nose straight down as fast as I could and accelerate away.

The 109 dives steeply, hurtling towards the ground:

> Bader: You see I'm behind him about to shoot him down and he does that . . .

The Spitfire pilot tries to give chase, sticks his own nose down in a dive to follow, but there's a problem:

> Bader: When I push the stick forward, all the petrol goes away from the carburettor and the airplane goes

pff_tt pffit pop bang – all this sort of banging and so on, and things come out from the bottom of the cockpit – you know, the spanners, that people left on the floor and so on, dust and stuff – so that's no good! [1]

In a steep dive, the Spitfire's engine splutters, and loses power, just when you need it most. The Spitfire can't handle the dive, and the German pilot knows it. He makes his escape.

Sholto Douglas, Deputy Chief of the Air Staff and later head of Fighter Command, is well aware of the problem: 'The engine defect of cutting at the commencement of a dive has been a sore point for some months, and has probably saved the lives of many enemy pilots who have traded on it.' [2]

So what's going on?

The answer lies deep inside the powerful Rolls-Royce Merlin engine.

This problem, the power cutting out in a steep dive, can be traced back to a small but essential part of the engine, called the carburettor.

The carburettor gets the fuel out of the petrol tank, and puts it into the engine. It's a little chamber about ten centimetres high, with a float that goes up and down. The float, when it's up, stops the fuel coming into the carburettor. The engine sucks fuel out from the bottom, the float comes down and more fuel comes in at the top.

This float system in the carburettor makes sure the engine always has just the right amount of fuel. But the problem is that it needs gravity in order to work properly. That's something you can take for granted in a car, but it gets a little less reliable in a mid-air dog-fight. You could be upside down, or racing towards the ground in a steep dive, and suddenly you're in 'negative g'. The fuel sloshes about, and the carburettor's clever little float simply stops working. One minute, the engine is starved of fuel; the next, it's flooded.

All of that means that the engine misfires, and loses power. Which is not what you want when you're chasing a German Messerschmitt across the Kentish sky.[3]

The pilots are well aware of the problem. Those who live long enough, like Douglas Bader, develop their own strategies to prevent the potentially deadly stall:

> Somehow we've got to keep the gravy in the carburettor you know, and so, we used to do a half roll, to keep the negative g, and then you roll round, and hope that you were in behind him. Usually of course he was way apart, running like a lamplighter you know – couldn't get him. Now, although our manoeuvre looks complicated, actually after a bit you got frightfully good at it you know. So you could nip around and he was still there, but a bit further away.[4]

That works for Douglas Bader, but you can't have your top pilots having to flip upside down every time they want to dive. This problem needs fixing, and the Ministry of Aircraft Production challenges two organisations to solve it – the engine manufacturers themselves at Rolls-Royce and the best aeronautic researchers in the country, the team at the Royal Aircraft Establishment. The RAE is a place where the finest minds in British engineering devote themselves to perfecting every part of Britain's aircraft, making them faster, more powerful, more dynamic in the air.

In 1940, the Spitfire problem landed on the desk of the Head of the Carburettor section at RAE, probably the person in the country best equipped to solve this precise problem. In 1940, that job was occupied by a fairly extraordinary person. Standing 5′1″ inch tall, with thick glasses, and a stern expression behind which the occasional flicker of wicked humour could be seen. The extraordinary thing, even in wartime, was: she was a woman – a woman always addressed by her staff as Miss Shilling.

Beatrice Shilling was the daughter of a butcher from Kingston upon Thames. From a very young age, it was clear that Bea, as the family called her, was different from other middle-class Edwardian girls.

> As a child I played with Meccano, and spent my pocket money on penknives, a gluepot, an adjustable spanner and other such simple hand tools. When broadcasting started from the 2LO stations in London I built wireless receiving sets. When I was sixteen years old I bought a small second-hand two-stroke motor-cycle. I got a lot of pleasure from dismantling and rebuilding this motorcycle, and my parents recognised that I had some mechanical skill.[5]

Beatrice's mother encouraged her hobbies, but the route from an interest in mechanics to an engineering profession was largely untrodden by women. During the First World War, thousands of women had carved out roles for themselves in heavy manufacturing and engineering, but when peace came there was enormous pressure on them from both management and the trades unions to abandon their lathes and workbenches and make space for the men returning from the trenches.

The Women's Engineering Society was set up to support the stubborn few who were determined to continue making a living from their new skills. The Society's members were keen to inspire the next generation, so a letter was sent to a number of schools, including Dorking High, where Beatrice was a pupil, offering bright girls the marvellous opportunity of a three-year apprenticeship in electrical engineering.

With a gentle but determined shove from her mother, Beatrice applied. She was selected for interview and met and impressed Miss Margaret Partridge, owner of the Exe Valley Electricity Company. At just seventeen years old, Beatrice left Surrey and took the train to Devon to learn a trade and play her part in the electrification of rural England.

She quickly mastered wiring up homes and the maintenance of the primitive diesel engines that produced the electricity. Miss Partridge was very pleased with her protégée:

> Beatrice Shilling is going to do very well indeed. She has taken over a big power station plan to do – plan and projection – and set it all out and traced it with a real engineer's understanding. Not traced only but carried out layout, and all. Also, the boys like her. They couldn't stand Miss Shatner and softly and silently managed to get rid of her whilst I was away in Leeds.[6]

The pilots of Britain's Spitfires and Hurricanes would soon have cause to be grateful for Beatrice's engineering skills and her ability to get the best out of her male colleagues.

Miss Partridge recognised that Beatrice had ambitions beyond the hedges and lanes of Devon and encouraged her to apply for a place on a degree course. She was accepted by the Department of Electrical Engineering at Victoria University of Manchester. No woman had studied engineering at Manchester before her, but she was very pleasantly surprised to find that another brave young woman, seventeen-year-old Sheila McGuffie, had also registered as a first-year student:

> We two women students were a slight embarrassment to the Engineering Department at times. On a boiler trial in hot weather, for example, the normal state of undress of the stokers was rapidly abandoned when it was remembered that women were present.[7]

Alongside her studies, Bea carried on tinkering with motorbikes. She was never happier than with her head buried deep in a deconstructed engine, spread out on the living room floor. Her grand-niece, Jo Denbury, remembers that it was a habit that continued well into retirement.

> I suppose my earliest memory of her was probably going to her house, I think she was about 60. You'd go in through the side because the sitting room was turned into a kind of mechanic's workshop. I always remember the smell, which was the smell of oil. And in there would be bits of engines and carburettors and you know vices on the side of a table where something was being worked on. But it wasn't really your standard front room!
>
> The excitement was always about the car, how's the engine running, oh let me have a look under the

bonnet, scoot underneath the car, have a check. And it was always the women. My mother, and Aunty Bea as she was known, would always have their heads in the bonnet of a car.[8]

At university Beatrice took part in motorcycle trials in the Peak District, but it was after graduation and her return to Surrey that she seized the chance to take up professional racing. With three miles of concrete track and steeply banked corners to maximise speed, Brooklands was the home of motorcycle racing in Britain. On race days it was packed full of like-minded enthusiasts and petrol-heads. For Bea this was heaven on earth:

> As soon as women were permitted to race motor-cycles at the Brooklands race track, I started to look at motorcycles which were suitable for racing. I bought a demonstration model 500cc International Norton. I modified the engine, increased the inlet valve and port diameters, and was placed several times, winning one race from scratch![9]

Brooklands offered generous cash prizes for race winners, but the award that was truly coveted was the Brooklands Gold Star, presented to anyone who could reach the then incredible speed of 100 miles per hour. In 1934, on a hot, sticky August race day, Beatrice was competing for just the second time at Brooklands. Her slim, angular figure, head to toe in black leather, immediately caught the eye of the reporter from *Motorcycling* magazine:

> A feature of the first handicap was the brilliant riding of Miss B. Shilling on a very standard-looking 490cc Norton. After a slowish first lap, she made up for lost

time with a second circuit of 101.02 mph, thus joining the select ranks of Gold Star holders, being the
second woman motorcycle racer to do so.[10]

Between races, Bea was job hunting. Prospects, though,
were few and far between. There were long waits in
company corridors only to be laughed at or dismissed,
there were letters of rejection addressed formulaically to
'Dear Sir'. One interviewer even disparaged her racing
record, sneering that the men probably let her win. It was
a dispiriting time.

But Bea's determination paid off. In 1936 she secured a
junior position at the Royal Aircraft Establishment. It was
to be the start of a distinguished thirty-five-year career at
the heart of Britain's world-leading aeronautics industry:

> I was offered a job as an assistant in the Technical
> Publications Department. The Air Ministry, having
> had some experience of women's work in the First
> World War, were not entirely unsympathetic. I spent
> the next eight months or so writing aero-engine
> handbooks. I then managed to get transferred to the
> Engine Department, where I found the familiarity
> with the detailed construction of several engines I had
> acquired raised my standing in the workshops.[11]

It was at the RAE that she bumped into a smiling,
strikingly handsome mathematician in the Mechanical
Test department, George Naylor. Well over 6 feet, to her
5´1″, he also raced bikes, although he wasn't quite as fast
as Bea.

George was smitten, but Beatrice believed in a marriage of equals. She told him that, once he got a Gold Star

of his own, she would marry him, and not before. It took him two years of time trials, but in 1938 he flew round the Brooklands track, winning himself the coveted Gold Star, and an even more coveted new wife.

When the war pulled them apart, George flying Lancaster bombers and Beatrice problem-solving for Spitfires, they wrote to each other often, as couples do, about everything and nothing. Beatrice offered her novel solution to an infestation of spiders: 'There are plenty of pockets of resistance in this house, so I decided a flame-thrower was the only thing for under the sink.' She also reminisced about her childhood:

> I was ticked off many times a day at school. I used to look so miserable that they used to apologise for talking so harshly. If you could establish that if you got ticked off you burst into tears, you could get away with anything.

And they shared the tough challenges of their war work:

> The position at work is very bad, I am sadly behind. I have three jobs I want flown and I can't get them done without going and urging them myself. Still, I'll finish sometime . . . We're trying to get extra fuel through a carb and have it ready for fitting to an air-craft at 8am tomorrow. Needless to say it is 11pm and some silly bloke has switched everything off because I took the troops out for some beer at 9.45.

In among the daily updates, there's the occasional lover's tiff and hints of the pain of wartime separation faced by so many young couples:

Darling, it is a shocking business this being separated. I haven't seen you for eighteen days, it is too long . . . you know dear that I am far too much in love with you to waste a three minute phone call being annoyed about you going to a dance or something. All my love, darling B.[12]

By the time the Spitfire engine problem was passed to the RAE, Beatrice was a well-respected engineer, managing a small team of carburettor specialists. From the outbreak of war, they had been concentrating on the problem of cold starts. In the harsh French winter of 1939–40, frustrated Hurricane pilots had watched as German bombers flew unchallenged over their airfields, completely unable to start their engines. It was a headache that disappeared as the spring of 1940 turned into the warmth of the Battle of Britain summer, replaced in her team's attention by the tricky conundrum of those stalling Spitfires.

The engine cuts had been observed in flight tests of the Spitfire as far back as 1938, but the assumption had been that Britain's fighters would be tackling high-altitude bombers. The spectacular aerial circus of the dogfights with Messerschmitt 109s had not been anticipated. The reports that Beatrice received from the front line were certainly concerning. Wing Commander Bob Doe described his first experience of the problem when he flew into turbulence caused by the slipstream from the Messerschmitt 109 he was attacking:

I opened fire at about 300 yards and seemed to hit the one I was aiming at because he pulled up sharply. At this moment (I must have been down to about a hundred yards), I hit his slipstream and my engine cut – stone dead![13]

The source of the problem, it seemed, was the SU AVT/40 carburettor fitted to the Merlin engines used in Spitfires and Hurricanes. Beatrice set up a test-bed, with rigs, and a hydraulic fuel pump. She somehow procured a Fairey Battle light bomber and conducted in-flight tests on its Merlin engine, tinkering with fuel pressure settings and the steepness of dives.

The problem proved fiendishly complex. For nineteen hours a day her team tested, prodded and poked at possible solutions. The complexity of the cutting-edge aero engine meant that there were many factors to consider. Fuel pressure and temperature, altitude, the steepness of the dive – all had to be accounted for.

Winter had returned before the exact nature of the problem was uncovered. Solutions were attempted, both by Beatrice's team and their friendly rivals at Rolls-Royce. Some succeeded in solving the first part of the problem – the loss of fuel to the engine under negative gravity – but nothing seemed to prevent the next stage, the flooding of the engine with an excess of fuel when the aircraft pulled out of its dive.

In the early spring of 1941, Beatrice finally had her answer. It was astonishingly simple, but the engineering know-how behind it was tremendous. It was just a tiny washer, a metal ring no bigger than a fingernail, with a carefully chamfered inner edge. Fitted between the end of the fuel intake pipe and the entrance to the carburettor fuel inlet gallery, it allowed enough petrol through for maximum power but not enough to flood the engine. The solution – the brass restrictor – was unveiled at the Derby works of Rolls-Royce in February 1941.

A jubilant Sholto Douglas, now head of Fighter Command, promised Beatrice's team that, 'If we can get your modification into our squadrons quickly we shall unquestionably spring a nasty surprise.'[14]

Sir Stanley Hooker, legendary mathematician and later a jet engine pioneer, had been leading Rolls-Royce's own efforts to find a solution. Speaking after the war, he seemed a little less than magnanimous at being beaten by a woman:

> It was cured, in a very simple way by a famous lady engineer from the Royal Aircraft Establishment, Miss Shilling, and she put a small thing about the size of a penny with a hole in it and we called it Miss Shilling's Orifice, if I remember at the time.[15]

The washer is more properly called the RAE Restrictor. But the joke name, the bawdy double entendre, really stuck. Miss Shilling's Orifice was what it became known as. Although, never to her face.

Crucially, the Orifice could be fitted without removing the engine, or even the carburettor, from the plane. There was certainly no time to be lost – the washer had to be fitted into every Spitfire in the country. Beatrice wasn't about to trust something so important to anyone else. She delivered the restrictors herself, in her usual style – at enormous speed. On her trusty Norton 490cc, with a bag of tools on the back, she distributed her sack full of Orifices to the mechanics at RAF bases around the country.

The RAE restrictor did the job. The dogfighting pilots risking their lives thousands of feet up could swoop and dive with no loss of power to the engine, giving them a vital edge in so many of the great aerial battles of the war.

Beatrice's great-niece, Jo Denbury, is determined that her vital role in the success of the Spitfire should be remembered:

> Bea believed in 100% equality. She just didn't think gender mattered, when it came to her work. But she did very much want it to be acknowledged that she had done what she had done, and she did campaign for more acknowledgement and she campaigned for equal pay too. She didn't have children, unfortunately, so it's fantastic that she's being rediscovered. There's now a scholarship to Manchester University in her name.[16]

After the war, Beatrice often ruffled the feathers of RAE management with her blunt approach, but she continued to work on some of the British government's top aviation priorities – guided missiles, observation aircraft and the Blue Streak, Britain's attempt to build its own ballistic missiles. By the time of her retirement in 1969, she was aware of the doors she – and other wartime women – had opened. But she was also aware just how much there was still to do:

> Many women were recruited by the Royal Aircraft Establishment and other government laboratories and workshops during the Second World War. By the end of the war, practically all posts in the Scientific Civil Service were open to single women. Equal pay, introduced over seven years, followed. Today girls are accepted for both craft and college apprenticeships in

government establishments. Prejudice against women is dying fast in Britain. However, while women engineers have won parity with men, there is a fight for all professional engineers in Britain if the present excellence of our engineering is to be maintained.[17]

*

Beatrice Shilling had achieved a remarkable feat in resolving a fundamental problem of the Spitfire's design, but niggling new issues continued to emerge. As the war progressed and the slim Spitfire chassis was loaded with more powerful engines and heavier weaponry, aircraft began to fall from the skies. Solving this mystery would require an air investigator with a deep knowledge of the Spitfire.

As the pilot credited with firing the first shots of the Battle of Britain, Wing Commander John Peel was one of the most experienced combat veterans in the Royal Air Force. He'd chalked up a fistful of aerial victories and been shot down three times, once rescued under heavy fire by a naval patrol boat just metres from the French coast.

On a routine patrol in a Spitfire Mark V, Peel lost control while pulling out from a high-speed dive, causing the g-force to suddenly drain the blood supply from his head. He collapsed, keeled forward and his head hit the control stick. The Spitfire accelerated toward the ground. At the last possible moment Peel came to and somehow wrestled the plane level. He landed safely but was deeply shaken by his Spitfire's devilish handling.

A new arrival at Peel's 65 Squadron was Tony Bartley. He'd been flying Spitfires since Dunkirk, most recently working as a flying instructor, a film actor and as a production test pilot with Supermarine. He was now back on front-line duty with this rather jumpy group of pilots. Peel's near-death experience seemed to confirm a growing suspicion in the squadron. There's something wrong with these Spitfires.

Bartley knew just who to call. Jeffrey Quill, Supermarine's Chief Test Pilot, had been receiving more and more reports of Spitfires failing in mid-air. Several pilots had been lost, far from enemy action. The Ministry of Aircraft Production, Supermarine's engineers and Quill himself were all completely stumped. Were these random cases of pilot error or a consequence of the increased demands being made of the Spitfire chassis? Now at least Quill had a suspect plane and a surviving pilot. He could begin the search for clues.[18]

Each new marque of Spitfire was loaded with more power and more weight. Supermarine engineers were convinced that Mitchell's delicate frame could sustain the pace of development, but, through 1942, evidence had begun to build that some aircraft were suffering fatal cases of structural failure.

After the near-death experience of Wing Commander John Peel, Quill flew to 65 Squadron's base at Debden in Essex to investigate. He took Peel's aircraft up and, sure enough, it felt dangerously unstable. This was not a case of pilot error. He tried three more planes from the squadron. Again it was clear that something was compromising their stability. The centre of gravity of all Spitfires depended upon the load of fuel, armaments, ammunition and other equipment carried. Supermarine provided users with a set of loading rules designed to maintain a stable centre of gravity for each flight, but in the heat of battle these instructions were patchily applied or ignored completely.

It was just too dangerous to rely on hard-pressed squadron mechanics following the complex instructions to the letter. A factory solution was needed.

Back in Hampshire, Quill helped the engineers test the use of counter-weights to balance the aircraft, but these could only be an interim measure. The breakthrough came when Quill bypassed his own colleagues and raised the problem with friends working at the Westland Aircraft company in Yeovil in Somerset. They already had lots of experience with the Spitfire, being responsible for building the naval variant of the plane, the Seafire. Their idea of adding a bulge to the elevator – the horizontal flaps at the rear of the Spitfire – was further developed by Supermarine's engineers at Hursley Park and fitted to new aircraft.

Once installed, the mid-air break-ups stopped and Spitfire pilots could focus their attention on the Luftwaffe without the distracting worry of a mysterious flaw in their aircraft.

14

The Spitfire Becomes the Star

The flaws in the Spitfire wouldn't become public knowledge until well after the war. Since the runaway success of the Spitfire Fund, it had been very apparent to the Ministry of Aircraft Production and their colleagues in the Ministry of Information that the Spitfire was propaganda gold. The performance of this beautiful machine in the Battle of Britain had captured hearts at home and abroad. As a positive tool to maintain and expand support for the Allied war effort, it was incomparable, and the British government was determined to wring every ounce of value out of the growing legend of the Spitfire.

It was late November 1941 when three of the RAF's finest young Spitfire pilots met at an airfield in southwest England. Tony Bartley, Brian Kingcombe and Christopher 'Bunny' Currant were the very epitome of Churchill's 'Few': public schoolboys of charm and wit. All three fought in the desperate battle for France, all three ravaged the Luftwaffe in the Battle of Britain.

They were here, of course, to fly Spitfires, but this particular week was a little different from the rest of their war. For one thing, there was very little risk of encountering a Messerschmitt; in fact, it felt very much like a well-deserved holiday for the three aces. No fighting, soft hotel beds, convivial evenings in the bar and fragrant young women offering them tea and cake whenever they took a break from filming. Bartley, Kingcombe and Currant were on set to play themselves in a film, *The First of the Few*.

The Battle of Britain, with its heroes and villains, action and romance, was an obvious subject for the cinema. Long before the Luftwaffe had given up its daylight raids, writers and producers were hawking stories around the movie studios of Hollywood and the Home Counties. The Crown Film Unit had already produced a popular docudrama based around the lives of the ferry pilots taking Spitfires from the factory to the front. Hollywood just loved the stories of American pilots defying their own government to join the RAF and save Britain, shooting three very similar films just as the United States entered the war.

All three featured a handsome but slightly arrogant American pilot discovering humility and the love of a good woman in the heat of battle. *International Squadron*, starring the future US President, Ronald Reagan, was first off the blocks in late 1941. It was thin stuff, featuring aircraft that, to a British audience, clearly had nothing to do with the RAF or the Luftwaffe. Betty Grable and Tyrone Power added some star power to Daryl Zanuck's popular *A Yank in the RAF*, while *Eagle Squadron*, released in August 1942, was initially intended as a documentary featuring real American pilots in Britain. It benefited from some genuine combat footage supplied by the Ministry of Information, but the critic at the *Monthly Film Bulletin* was unimpressed. This, he complained, was 'the mixture as before'.[1]

What British cinema audiences really craved was a home-grown telling of their story. That was precisely what the team behind *The First of the Few* intended to give them. The Australian writer Henry C. James had been quick to note the success of the Spitfire Fund and the reverence in which the name of R. J. Mitchell was held in aviation circles. He contacted Mitchell's widow, Florence, and enlisted her help in writing – and radically romanticising – his story. Producers George King and John Stafford liked the script and purchased the option to film it. With their own money now on the line, King and Stafford were anxious to get filming underway, informing the eager readers of *Kine Weekly* that an exciting new feature was on its way:

> He died that England might live, and England lives because he died. Reginald Mitchell, inventor of the Spitfire, Englishman, visionary and patriot, whose genius rides the skies of England, and whose winged avengers tear down the Goering gangsters from the blue, and send them crashing to the earth of mother England, which he loved.[2]

King and Stafford, it's no surprise to learn, were small fish in the small pond of the British film industry. Luckily for them a much bigger fish was prowling the same waters in search of inspiration. Leslie Howard was fresh from playing Ashley Wilkes opposite Vivien Leigh in *Gone With the Wind*, the world's highest grossing movie. With his long face and deep-set eyes, he had been inducing swoons in the more sensitive of film fans since the 1920s, but Howard was a movie idol who hated the Hollywood machine. He was a film star desperate to take on a more challenging role than the handsome hero. *The First of the*

Few offered him the chance to co-produce, direct and play the lead.

The producers approached Winston Churchill for permission to use an extract from his famous Battle of Britain speech as the title. Somewhat flattered, he rang the head of Fighter Command to strongly suggest that all possible assistance be granted to the production. Howard was offered the loan of a Blenheim bomber, a captured Heinkel 111 and the Spitfires of 501 Squadron for the aerial action sequences. As well as the military hardware, Howard's star power attracted two top names of the day. Matinee idol David Niven was released from his army regiment for five months (or just four weeks, if you believe Niven's brilliantly imaginative autobiography, which you certainly shouldn't) and William Walton agreed to compose the score. With Leslie Howard's daughter and his mistress both added to the cast, he was ready to start shooting.

The film begins with a newsreel-style summary of Nazi beastliness before we join the aftermath of a German bombing raid on England. Spitfires land, fresh from the fight, and our genuine young combat veterans trade period banter with actors playing other pilots and ground crew:

> 'Hello Bunny, how'd you get on?'
> 'I got an 88 and had a crack at a Dornier.'
> 'Good show, that's grand.'

Once the day's fighting is done and a lost comrade briefly mourned, the pilots gather to confide in their commanding officer, played by David Niven. They wonder aloud about the origin of their beloved Spitfires. The Wing Commander reveals that he played a part in the true story of the plane's genesis, as a good friend of R. J. Mitchell.

We now meet Mitchell himself, portrayed by Leslie Howard as an upper-class dreamer rather than a short-tempered middle-class engineer. Basking with his wife on a lonely clifftop, he watches the seabirds dart and weave and imagines an aeroplane that can rival their ease and confidence in the air. This is a portrait of the inventor as a troubled aesthete, determined to finish his life's work before his diseased body (ailment politely unspecified) gives up the fight.

The movie is the perfect complement to the productions of the Ministry of Information and the Crown Picture Unit. Where those films stress the collective battle against Nazism, the grit and pluck of the ordinary worker doing his or her bit, *The First of the Few*, from its title onwards, presents the Spitfire as the creation of one extraordinary man. While films like *Listen to Britain* and the Blitz docudrama, *Fires Were Started* focus on the urban experience of war, *The First of the Few* positions Mitchell in a thoroughly rural version of Olde England, a vision perhaps more appealing to American movie-goers.

The film opened on 30 August 1942 at the Leicester Square Theatre. The public response was immediate and enthusiastic. It was the number-one draw at the box office in September 1942 and ended the year as the top British film. At a time when some 70% of the population regularly visited the cinema, there's no doubt that the message of *The First of the Few* reached an enormous audience.

That message, as far as the film critics were concerned, was loud and clear. In *The Spectator*, Edgar Anstey wrote that: 'The hero of *The First of the Few* is the Spitfire and the film contrives to invest this shining creature with all the historical monumentalism of HMS *Victory*.'[3]

The plane's iconic status was confirmed soon afterwards in one of the most beautiful single shots of 1940s cinema. At the opening of Powell and Pressburger's *A Canterbury Tale*, the hunting hawk unleashed by one of Chaucer's pilgrims transforms itself into a Spitfire roaring overhead. The Spitfire is a symbol of modernity, but also the bridge between two eras of British greatness and decency.

Buoyed by its British reception, the producers' expectations ran high for the US release of *The First of the Few*. *Mrs Miniver* – the Oscar-winning story of an ordinary British housewife coping with the privations of war – was still packing cinemas coast to coast, and Leslie Howard was a hugely popular star, certain to attract an audience. The Hollywood film producer Samuel Goldwyn had given permission for one of his top stars, David Niven, to appear in this independent picture in exchange for the US rights. Goldwyn's team were ready to push the film hard, with a promotional campaign led by posters promising 'The Story of the Plane That Busted the Blitz'.

The First of the Few did conquer America, but Leslie Howard wasn't there for the premiere. Just two days before the film's US release, he went missing.

Unlike many of the pampered Brit Pack stars in Hollywood, Leslie Howard had rushed home at the outbreak of war. He was eager to do what he could for the war effort, making the most of his star power and contacts to produce a series of films designed to prop up morale at home and encourage the United States to support Britain.

In May 1943, Howard was touring Spain and Portugal at the request of the British Council. His lectures were purportedly in support of his latest film, a medical drama called *The Lamp Still Burns*, but there's little doubt that his presence helped open doors for diplomatic contacts. The Spanish Fascist dictator General Franco was said to be a passionate fan of *Gone With the Wind*. Late on the afternoon of 31 May, Howard and his friend, accountant and sometime agent, Alfred Chenhalls, booked themselves on to a BOAC flight from Lisbon to Whitchurch, near Bristol. Both the British and German governments had agreed to respect Portuguese neutrality. This allowed for regular scheduled flights between Lisbon and the UK, which soon became notorious for carrying secret agents and escaped prisoners of war. That reputation inevitably attracted German intelligence operatives keen to scan the passenger lists. The Lisbon terminal at Portela to the north of the city was described by BOAC's local operations officer as 'like Casablanca but twentyfold'.

Since 1942, the supposedly safe flight path over the Bay of Biscay had become busier with military aircraft from both sides. The scheduled BOAC Lisbon-to-Bristol flight was attacked on two occasions by German fighters, but each time the Douglas DC-3 was able to limp home. Flight 777-A took off early on the morning of 1 June 1943 with four Dutch aircrew and thirteen passengers on board. Alongside Howard and Chenalls, there was an oil executive, a prominent Jewish activist, a Reuters journalist and an inspector of British consulates. As the flight turned north-west from the coast of Spain, it was set upon by a group of Junkers Ju88 maritime fighters. The wireless operator on-board reported the attack to Whitchurch control:

> I am being followed by strange aircraft. Putting on best speed . . . we are being attacked. Cannon shells and tracers are going through the fuselage. Wave-hopping and doing my best.[4]

On 3 June, the *New York Times* reported: 'A British Overseas Airways transport plane, with the actor Leslie Howard reported among its 13 passengers, was officially declared overdue and presumed lost today.'[5]

Within days of the attack, rumours began to spread about the motive for this unprovoked attack on a civilian airliner flying a safe passage route. Several of the passengers might have been of interest to the Germans, but the first theory to gain a following was that Nazi intelligence believed Winston Churchill to be on board the flight. Churchill had been in North Africa for talks with US General Dwight D Eisenhower and it was quite feasible that he may have chosen to travel home via Lisbon. The passing resemblance of the stocky, cigar-chewing Alfred Chenalls to Churchill added fuel to this theory.

In his 1984 book, *In Search of My Father*, Leslie Howard's son, Ronald, shared his belief that his father had been the real target. Howard's films, such as *Pimpernel Smith*, had been invaluable in the propaganda war with the Nazis and there were rumours that his activities in Spain went beyond the promotion of his films and British values. An ex-girlfriend was said to be married to a prominent member of Franco's Falangist party. Was he perhaps engaged in talks with the Spanish dictator, trying to persuade General Franco to turn away from the Axis powers? The fevered atmosphere of conspiracy around Flight 777 threw up one final rumour. Could German agents have been so taken in by *The First of the Few*, which was apparently screening at the time in Portugal, that they mistook Howard for R. J. Mitchell? To assassinate the inventor of Britain's greatest war weapon would indeed have been a military triumph, had Mitchell not already been in his grave for five years.

The truth is almost certainly more prosaic. Post-war interviews with surviving Luftwaffe crew from the attacking Junkers 88s suggest that visibility was poor that morning. In the cloudy sky, the grey silhouette of the airliner was identified as an enemy craft. The plane was already fatally wounded before the neutral airline markings were spotted and the attack called off. The Luftwaffe veterans claimed to have no knowledge of the passengers on board.

Whatever the truth, a true screen legend lay beneath the waves of the Bay of Biscay, his mysterious death adding another layer of intrigue to the Spitfire myth.

The profile of *The First of the Few* was boosted by the death, at the hands of the Nazis, of one of cinema's favourite stars. Even the most imaginative of Hollywood screenwriters would have found it hard to conjure up such a propaganda boost for the Allied cause. The Spitfire's status as the breakout star of World War Two was assured. When *The First of the Few* was released in America, it was re-titled, simply, *Spitfire*.

The romantic image of the Spitfire was now set in stone, but its reputation as the best fighter of the war was under threat as never before. While movie-makers and propagandists were busy turning it into an icon, Germany's engineers were building something to blast the Spitfire from Europe's skies.[6]

15

A Deadly New Foe

It's November 1941 when Squadron Leader Johnnie Johnson spots an unusual shape in the distance: 'We were puzzled by the unfamiliar silhouettes of some of the enemy fighters, which seemed to have squarer wing tips and more tapering fuselages than the Messerschmitts we usually encountered.' A barrage of machine gun and cannon fire blazes from the wings and nose of the fast-approaching aircraft. Johnson takes evasive action in the nick of time.

> Whatever these strange fighters were, they gave us a hard time of it. They seemed faster in a zoom climb than the Spitfire, far more stable in a vertical dive, and they turned better than the Messerschmitt, for we all had our work cut out to shake them off.[1]

In the months since the Battle of Britain, RAF Fighter Command had been operating an offensive policy, designed to tempt German fighters into the air, where they could be attacked by large numbers of British fighters.

Flying the latest Mark V Spitfires, pilots felt confident of a slim combat advantage, even against the latest upgraded Messerschmitt 109Fs. That confidence dissolved in the autumn of 1941.

The existence of a new German fighter had been rumoured for months. A captured German serviceman had revealed the basics of the new aircraft in early 1941 and, by 13 August, the Air Ministry's Weekly Intelligence Summary was sufficiently confident to report that this fighter was a low wing monoplane, with a two-bank radial engine. The report estimated its top speed at 370–390 mph.

Those bare facts alone were enough to alarm the Air Ministry; the reality experienced by Johnnie Johnson revealed a machine with the potential to turn the tide of war.

Kurt Tank, the designer of this astonishing aircraft, was an ex-cavalry officer. He was impressed by the Spitfire and the Messerschmitt 109, with their speed and elegance, but he thought the Luftwaffe needed something tougher, a cavalry charger rather than a racehorse. His first prototypes were tested before the outbreak of war but it took another two years of balancing the aircraft's weight with its wing size before the Focke-Wulf 190 A1 was ready to savage the Royal Air Force.

Early encounters proved without doubt that the Spitfire Mark V was comprehensively outclassed by the Focke-Wulf 190 and Fighter Command was forced to rein back its offensive operations in France, Belgium and Holland. Without mastery of the skies, invasion of the Continent was unthinkable and Britain itself might once again be threatened. The secrets of the Focke-Wulf had to be uncovered. A combined forces operation was ordered – the Army, Navy and Air Force would work together on one of the most audacious plans of the war.

No. 12 Commando unit was stationed close to Jeffrey Quill's home in the Hampshire village of Bursledon. One day in 1942, the head of the Commando's E troop, Captain Philip Pinckney, dropped in uninvited for tea. Quill instantly recognised a very particular type of Englishman:

> Philip was a man of rare and timeless character. One might have encountered him accompanying Drake's raid on the Spanish treasure ships in Panama, or steering a fireship amongst the Armada anchored off Calais, or with Shackleton on his epic open-boat journey from Antarctica to South Georgia. Equally he was in no way out of place in the Ritz bar.[2]

In short, Pinckney was just the type of chap who built the British Empire – a man, as Quill put it, who was very unlikely to survive the war.

Pinckney had a plan. For it to succeed he absolutely needed Quill's special skills and he absolutely refused to take no for an answer. Within days he had Quill in training, paddling an inflatable boat on Southampton Water. The plan was to steal a Focke-Wulf.

A naval gunboat would ferry Pinckney and Quill – just the two of them – under cover of darkness to the French coast. They would row the last few hundred metres in an inflatable boat, bury it and proceed inland. Quill took up the story in a BBC documentary:

> We were going to be put ashore and get into a German airfield, probably Abbeville. I made the stipulation that the Army would get me into a cockpit of the 190 with the engine running and from then on it was up to me. I thought if we got that far I had a little better than a 50/50 chance of getting off the airfield and getting back.[3]

Quill felt reasonably sure that he could pilot a Focke-Wulf. As part of his work for Supermarine, he'd studied the instrumentation and control labelling in other German aircraft and he'd recently had the chance to fly a captured Messerschmitt 109, commenting rather sniffily that:

> I wish I'd had the opportunity to fly it before fighting against it because I'd have been much more confident. It had very heavy controls, it had a bad canopy and not a good view and it was nothing like as good an aeroplane as the Spitfire.[4]

It would be fascinating to see just how much better this new German wonder-plane was. The gung-ho commando was supremely confident that he could get Jeffrey to the plane, although it was never made entirely clear how he planned to get himself back to England. Whatever the gaping holes in the plan, HQ Combined Operations gave their stamp of approval and a top secret designation – Operation Air Thief. All Quill could do now was wait for the starting pistol. Nervously.

> One day Philip Pinckney arrived on an army motorbike and he had a very long face and he said, 'the whole thing's cancelled, some chap has landed an FW190 intact in Wales and the whole thing's off.' I said, 'I'm sorry, Philip', but I must say I was very relieved.[5]

Luftwaffe pilot Oberleutnant Armin Faber was part of a squadron of Focke-Wulfs that had pursued a raiding party of British Boston bombers back across the Channel. In a series of dogfights with the escorting Spitfires, Faber became disorientated and mistook the Bristol Channel for the English Channel. He reportedly gave a victory roll as he landed at what he assumed was a French airfield. It was, in fact, RAF Pembrey, a small air base in south-west Wales. The Sergeant on guard duty leapt on Faber's wing, pointed a flare gun at him and hollered, '*Hände hoch!*' before he had the chance to destroy the aircraft.

Within five days Faber's Focke-Wulf 190A-3 was being studied by Rolls-Royce engineers and Operation Air Thief was called off. Philip Pinckney survived his buccaneering lifestyle for another year before being killed in northern Italy leading a group of six SAS parachutists far behind enemy lines.

Quill, meanwhile, would soon find himself lining up against Oberleutnant Faber's Focke-Wulf, not in combat, but in a race that would decide the future of the Spitfire.

Extensive studies of the captured Focke-Wulf by the engineers of Rolls-Royce, RAE Farnborough and the Air Fighting Development Unit did nothing to ease the concerns of the top brass.

On 28 June 1942, Ernest Hives, head of aero engines at Rolls-Royce, wrote to Sir Wilfrid Freeman, Vice Chief of Air Staff: 'We feel very depressed about the fighter position, both in this country and in the USA and this has not improved since one of our own pilots examined the Fw190.'[6]

Government documents of the time reveal something close to a polite panic about Britain's loss of air superiority. Chief of Air Staff Sir Charles Portal wrote to Winston Churchill to tell him that the power of the Fw190 meant that 'it will be advisable to moderate our day offensive'. Essentially this meant that it was unsafe for British aircraft to operate in daylight over Occupied Europe. Worse still,

Britain itself could very soon be vulnerable once more. The Managing Director of Rolls-Royce wrote to Sir Archibald Sinclair, Secretary of State for Air, referring to an understanding that 'the government anticipate a second "Battle of Britain" in a few weeks' time.'

Sir Archibald reflected the general mood of desperation in a telegram of his own to John Jestyn Llewellin, Minister of Aircraft Production, on 21 July 1942:

> The Germans are having too much success in their development of specialized fighters. We are being left behind. A turning point in the war has been reached. Our mastery of the daylight air is threatened . . . the fighter aircraft of the German Air Force have a higher performance than the aircraft used by the R. A. F. today.[7]

How might Sir Archibald and the hard-pressed Spitfire pilots have felt if they had known the truth about enemy aircraft production? Both the Fw190 and the latest Messerschmitt 109s were handicapped by supply restrictions in Germany. It was virtually impossible to get hold of 100 Octane fuel, other fuel grades were of dangerously variable quality and components in new aircraft were being made with economy steel. This was a formulation that lowered the nickel content, resulting in a shorter lifespan for engines, more frequent in-air engine failure, and restrictions being placed on the way that pilots could fly. The Spitfire was being beaten over France, Belgium and the Netherlands by planes that weren't even operating at their peak efficiency. Britain had plenty of fuel, and Rolls-Royce and Supermarine still had access to the highest quality components. Despite these advantages, they still couldn't beat the Germans.

British aircraft manufacturers certainly weren't sitting on their hands. Some of the country's finest engineering minds at Hawker, de Havilland and Gloster were busy developing new fighters, but there was a sense in the Ministry that politicians, civil servants and the manufacturers had failed to grasp the urgency of the matter. A rather extraordinary race was devised to sweep away any hint of complacency.

The captured Focke-Wulf would fly head-to-head against the best of the next generation of RAF fighters – the Hawker Typhoon. To provide an appropriate point of comparison, Jeffrey Quill would fly a plane that the watching officials would all recognise – the Spitfire.

For Quill the reputation of his beloved Spitfire was at stake. He was personally convinced that the Spitfire could be further developed to combat the new threats from the Luftwaffe, but he was worried that the Air Ministry didn't share his confidence.

Rather than fly a standard production Spitfire to Farnborough for the race, Quill chose a test plane that he'd been particularly impressed by – the prototype Mk IV, DP845, powered by one of the new Rolls-Royce Griffon engines.

Hawker test pilot Kenneth Seth Smith was to fly the Typhoon, H. J. 'Willie' Wilson, the RAF's specialist pilot of captured aircraft, would take on the Fw190 and Quill the Spitfire. The three rival aircraft took off from Farnborough and flew ten miles west to the village of Odiham. They arranged themselves in an approximate line and, on Willie's signal, all three opened their throttles and sped back toward Farnborough. Halfway along the low altitude course, the Focke-Wulf developed engine problems. Quill shot ahead of the smoking German and beat the Typhoon to the line. He was quite delighted with himself: 'That certainly put the cat among the pigeons in front of a high level audience including C in C Fighter Command, Sholto Douglas.'[8]

Proof had been provided that an all-new fighter wasn't needed, just a better Spitfire.

*

The Hawker test pilot, Seth Smith, was killed just a few weeks later over Surrey when the tail of his Typhoon detached in flight. Willie Wilson survived the war and went on to break the world air speed record in November 1945, taking a Gloster Meteor jet to 606 mph.

Jeffrey Quill also went on to test the post-war generation of jet fighters, but so many years of flying unpressurised aircraft at extraordinary altitudes and speeds had affected his health. In June 1947 he was flying the prototype of Supermarine's new Attacker jet at a height of twelve kilometres or 40,000 feet. He passed out, returning to consciousness nine kilometres down. He landed safely but his days as a test pilot were over.

Quill, though, stayed in the aviation business, becoming one of the directors of Panavia, the British-German-Italian alliance that designed the Tornado, which became the forty-year workhorse of the RAF and saw action in both Gulf Wars, in Kosovo, Afghanistan and in Syria in 2018. Based at Panavia's headquarters in Munich, Quill met and discussed new designs with Wily Messerschmitt, inventor of the Bf 109 and Kurt Tanks, the designer of the brilliant Focke-Wulf 190 that had taken so many RAF lives.

In a fifty-year career, Jeffrey Quill flew more than ninety different types of aircraft and had a key role in the design of military planes right into the 1970s, but it was his role as the bridge between combat pilots and the Spitfire designers that made a fundamental difference to Britain's war effort against the Nazis. Without his remarkable ability to establish a meaningful rapport with the designers and engineers on the ground, the Spitfire could not have been incrementally improved right through the course of the Second World War.

16

Hursley Matches the Germans

In his race with the Focke-Wulf, Jeffrey Quill had proved a point that the Supermarine team had been pressing for a long time. As early as 1940, Joe Smith, the man who took over Mitchell's role as Chief Designer, had become convinced that the Spitfire was capable of extensive, perhaps infinite, revision and development. That confidence was based on solid evidence. Before the outbreak of war, Supermarine had planned an attempt on the world speed record held at the time by Howard Hughes's H-1 Racer. Changes to the airframe and the use of a specially tuned Merlin engine could, Smith's team thought, take the air speed of the Spitfire up to 425 mph, compared with the Mark I's 362 mph.

Whether that confidence, or anything approaching that speed, could be translated quickly into production aircraft that would beat the Focke-Wulf depended upon the design team now assembled at Hursley Park in Hampshire.

There were concerns after the bombing of the Supermarine factory that the dispersal would make rapid development of new Spitfire models difficult. One of the secrets of Supermarine's success at Woolston had been the close contact between the design department and the factory floor. New ideas could be discussed with the men making the parts. Short-run batches of new components could be built and tested, giving R. J. Mitchell and Joe Smith the opportunity to amend and improve their ideas. With production now spread across dozens of sites, this intimate relationship between creativity and metalwork was broken.

There was no going back to the Woolston shore, so Smith had to find fresh ways to bring the makers and modellers together, ways that would actually speed up development and give the Spitfire a fighting chance of keeping up with the Germans.

Smith had learned his engineering skills at Austin Motors in Longbridge, Birmingham, before moving south in 1921 to join Supermarine. His talent was swiftly spotted by Mitchell, who promoted him to Chief Draughtsman, and they worked together, the temperamental genius and the calm and methodical pragmatist, until Mitchell's death. It was under Smith's leadership that the inspired but complex Spitfire prototype was turned into an aircraft that could be mass produced, and it was Smith who would lead on design matters after the dispersal from the bombed factory.

One of his first decisions at Hursley was to form a dedicated team capable of creating machine-tooling and prototype aircraft in close proximity to the designers.

The new Experimental Department began work in the original Georgian stable yard that still exists at Hursley, split between various barns and out-buildings. From the air the yard was invisible, covered by an enormous hangar roof. It was here that metal alloys were stress-tested to make sure they were capable of coping with the increased power and speed of the latest variants and their engines.

Prototype fuselages could, with a little squeeze, be assembled directly onsite.

In 1942 a new, purpose-built Experimental Hangar was constructed close to the southern, Southampton entrance to the Hursley Park estate. Eighteen-year-old Apprentice Aircraft Engineer Ken Miles was one of those who had to work out how to move a prototype fuselage from the stable yard to the new hangar. The old stevedore arches were far too tight for any kind of transport vehicle, so Ken and his mates just had to lift it themselves:

> The fuselage was lifted on to our shoulders and we proceeded to walk the half-mile to our new home. Our way took us through the orchard and when we arrived the fuselage was full of apples.[1]

Complete prototypes could now be built, with engines fitted and tested before being loaded on to Queen Mary trailers – extra-long truck-beds – for transportation to Worthy Down or High Post airfields for final assembly and flight testing.

Once fully installed and settled at Hursley Park, the Supermarine Design Office developed into a formidable technology hub. One of the secrets of its success, an idea replicated at today's tech giants, was the transformation of employees from workers into team members. With decent food in hearty portions and plenty of leisure opportunities onsite, staff members were encouraged to base their lives around the company. Photographs from the time and copies of the in-house magazine reveal the success of that strategy. There were annual Design Office dinners featuring menus with caricatures of senior management and songs, dances and comic sketches performed by the staff. Fancy dress football matches pitching the mathematicians and aerodynamicists of the Technical Office against the draughtsmen of the Drawing Office kindled a keen but, usually, good-natured rivalry.

Draughtswoman Stella Broughton remembers works outings to the Winchester Guildhall for table tennis competitions and piano recitals and even a particularly exciting trip to dance to the biggest stars of the day – Glenn Miller and his Army Air Force Band. She helped organise a sports week for the staff and 5′2″ Stella was, literally, roped into a three-legged race partnered with her 6 foot tall friend Noel Mills:

> We had only a few minutes to practise but came in a close second. We were very pleased with our result but I am sure if we had been given prior notice and a little more time to adjust, we would have won.[2]

At the same time that Stella was enjoying Hursley Sports Week, her German counterparts were surviving on shrinking rations, dreading the prospect of a posting to the Eastern Front. Messerschmitt 109s were being assembled by slave labour and many of the best engineering brains were already buried in the snow and mud of Stalingrad or the rubble of Minsk. As the war went on, the value of a content and motivated workforce would gradually make itself felt.

The close-knit team at Hursley Park accelerated the development of the Spitfire, responding again and again to the demands of a truly international war and the best efforts of the German engineers.

The Mark V, designed at Woolston, was adapted at Hursley for use in the deserts of North Africa and the jungles of Burma. Crucially, the Mark IX was ready for testing by the spring of 1942. With a top speed to beat the Focke-Wulf, a more powerful Rolls-Royce Merlin 61 engine and a distinctive four-blade propeller, it was just what the RAF needed. When production was ramped up, it gave pilots over Occupied Europe the edge they had been sorely missing.

Evaluated against a Mk V in early 1942, the Air Fighting Development Unit test pilot was impressed:

The performance of the Spitfire IX is outstandingly better than the Spitfire V, especially at heights above 20,000 feet. On the level the Spitfire IX is considerably faster and its climb is exceptionally good. It will climb easily to 38,000 feet and when levelled off there can be made to climb in stages to 40,000 feet by building up speed on the level and a slight zoom. Its manoeuvrability is as good as the Spitfire V up to 30,000 feet, and above that is very much better. At 38,000 feet it is capable of a true speed of 368mph and is still able to manoeuvre well for fighting.[3]

But the Hursley team couldn't pause for even a moment. Despite Allied bombing and a chronic shortage of parts, German engineers were still making enormous strides in the development of jet fighters and, soon after D-Day, V1 flying bombs began to hit Britain – another sharp reminder that the Spitfire had to keep getting faster, keep hitting harder, keep flying higher.

At the outbreak of war, the Spitfire Mk I had a top speed of 362 mph (582 km/h). By 1946, the Spitfire Mk 24 was in service – able to fly at 454 mph (730 km/h) and reach a flying altitude two miles, or three kilometres, higher. Rolls-Royce brilliantly uprated the raw power of its engines, but it was the team at Hursley Park that made this extraordinary speed of development possible, adapting the light and elegant original design until it was around twice as heavy and twice as powerful as the prototype K5054. At Hursley new and better designs could travel from the drawing board, through mock-ups to full-size prototype aircraft, all on one site that the Germans were never able to find and destroy.

Many versions of the Spitfire were born at Hursley House – interceptors, ground attack aircraft, reconnaissance planes; the Seafire that could take off from aircraft carriers; and the Mark IX that could match and beat the best the Germans had. As D-Day came and the Allied forces pushed through France toward Germany, the Spitfire would be the master of the skies, destroying German fighters, strafing tanks and supporting the bombers ripping the heart out of the Nazi war effort.

The Spitfire would be the only fighter aircraft built at the start of the war that was still being produced and sent into front-line action as the atomic bombs fell on Hiroshima and Nagasaki. It outlived all of its competitors – all thanks to the likes of Stella Broughton and the designers, engineers and draughtsmen and -women of Hursley Park. Stella is rightly proud of her work:

I'm very pleased that I was able to use my talents in such a way. It gives me a little bit of pleasure to know that I was one of the people who helped during the war to produce the plane that really saved this country.[4]

The people of a shattered company and a shattered city had found a way to build the Spitfire against the odds, they'd found ways to keep developing their aeroplane, maintaining its position as the very best fighter plane of the Second World War. At each stage of the conflict the builders and designers provided the right tools for the job, but those planes had to be flown and battles won against the most powerful air force that the world had ever seen. Brave young British pilots were joined by the defeated ranks of Poles and Czechs, French and Belgians. American fliers defied their government to cross the Atlantic in search of adventure, and Canada, New Zealand and the Caribbean supplied the Royal Air Force with many more.

Each pilot would form a bond with their Spitfire as they learned its tricks, forgave its flaws and flew higher and faster than any of them had ever dreamt of.

Part Three

THE FLIERS

17

A Bullet Through the Sky

The moment the cockpit slides back the smell hits you – what Squadron Leader Mark Sugden describes as 'a delicious mix of fuel and paint thinners'. To a wartime pilot, it was a potent reminder that they'd squeezed themselves into a small metal box surrounded by high octane fuel. This isn't a modern aircraft. There's no fire management system, no ejector seat to get you out of trouble. To be trapped in a smoke-filled cockpit, desperately trying to release the hood, was a horror that many endured and a fear that kept many more awake at night.

Mark is our key to the minds and memories of the fliers we're about to meet. He pilots the Eurofighter Typhoon jet for the Royal Air Force. He's flown combat missions over Iraq and Syria – he knows how it feels to shoot and be shot at. Squadron Leader Sugden is also a Spitfire pilot. Every summer he leaves the world of modern warfare behind and steps back eighty years into one of the Spitfires of the Battle of Britain Memorial Flight. From their base here at RAF Coningsby in Lincolnshire, they perform at air shows and memorial fly-pasts all over the country, offering a spectacular tribute to the pilots and crew who won the Battle in 1940.

We're in a hangar surrounded by Spitfires in various states of undress. Outside there's a Lancaster bomber taxiing and our conversation is regularly interrupted by the roar of the Eurofighter Typhoon jets taking off from the runway next door.

These sheds and airstrips may be slotted between the dykes and drained fens of central Lincolnshire, but they're close to heaven for anyone raised on black-and-white war films, anyone who's ever had freshly-painted Airfix Hurricanes and Messerschmitts dogfighting above their bed. There's a Dakota transport plane and a tractor ready to drag it through the enormous hangar doors, Merlin engines stripped for repair with plastic bags of screws labelled and tied to copper pipes. There are propellers propped against the walls and a Chipmunk training aircraft in bits on the floor, like a jigsaw puzzle to while away eternity.

Together we're peering into the cockpit of P7350. It's a Mark IIa, the only Spitfire left that flew in the Battle of Britain, and still takes to the skies today. One of the first aircraft built at Castle Bromwich, she was saved from scrappage and starred alongside Laurence Olivier and Michael Caine in the 1969 film epic *The Battle of Britain*.[1]

We're both silenced for a moment as a Typhoon roars down the runway. Mark pauses a little longer and takes a deep breath. When he flies this plane, he says, he's always aware that he's not the only one in the cockpit. He flies with the ghosts of the men that took her into war. Even now, after dozens of hours of flying this and the BBMF's other Spitfires, each flight is an assault on his emotions.

Mark's due to fly in an hour, off to the Clacton Air Show in the Flight's Mk IX Spitfire, so checks have to be made. The instrument panel and much of the cockpit layout is the same as you'll find on the Hurricane or the Mosquito. Pilots had to be able to switch smoothly from aircraft to aircraft, so there's a surprising consistency between the controls.

On the right are the engine instruments – oil temperature and pressure, RPM and fuel gauge. To start the Merlin there's a plunger – a syringe that primes the engine. It puts fuel directly into the manifold that supplies the mixture of fuel and air to the cylinders. When you see those stirring movie images of Spitfires belching flames before take-off, it's the sign of a heavy-handed pilot. He or she has put too much fuel into the engine, causing a damaging ignition in the exhaust stacks.

For display flying, some of the original controls are redundant. Mark's not allowed to take the aircraft above three kilometres – or 10,000 feet on the vintage altimeter – so there's no need for the oxygen supply, and with modern radio communications fitted, he won't be needing the Morse code tapper. That's just as well, as Mark's not sure how much Morse he can remember. Perched on the spade-grip control stick there's the weapons control toggle switch. There's a trigger for the 0.303-inch Browning machine guns on the top and a trigger for the 20mm Hispano-Suiza cannons on the bottom. If you needed to fire both, you would press the middle of the toggle.

Close to the cockpit floor are the rudder pedals. In the Eurofighter Typhoon, they're nicknamed 'footrests', as modern jets don't need a lot of manual steering. On the Spitfire, though, you have to work the rudder hard. Mark says that an hour in the Spitfire can make you feel like you've run a marathon. Your feet are working in the air to keep the aircraft balanced and, on landing, you need to 'dance on the rudders', making constant inputs to keep the plane straight.

A mildly misogynist wartime cliché went that the Spitfire was 'an angel in the air and a bitch on the ground'. From the cockpit, it's easy to understand the phrase. The powerful Merlin engine and the main fuel tanks are all in the nose, making it appear incredibly long. On the ground the plane is resting on its rear wheel, tilting the cockpit upwards so it's very hard to see anything useful past the nose as you taxi for take-off. Worse still, if you're not careful when landing, the tail can overtake the nose. Mark compares it to throwing a dart backwards – when the tail wheel hits the ground, the centre of gravity of the aircraft shifts dramatically.

Time now for the final pre-flight checks. Mark straps himself in, checks the flight controls aren't restricted, sets up the control surfaces for take-off and confirms the instruments are operating correctly. He primes the engine with a combination of the syringe and an electric fuel pump. Once the control tower confirms that the Memorial Flight formation is ready for take-off, I make myself scarce and Mark presses the start buttons beneath the instrument panel. He flicks on the magneto to supply current to the ignition system and the engine bursts into noisy, glorious life.

Spitfires are designed to get airborne quickly. Keep them on the ground too long and the powerful Merlin engine will overheat, so ground runs are kept short and sharp. Mark accelerates and pushes forward on the control stick. This lifts the tail off the ground, helps the aircraft gain speed and gives the pilot a better view of the runway ahead. Once he reaches take-off speed, Mark pulls gently back on the control column, leaves the ground and raises the undercarriage. The plane completes its transformation from clumsy, earth-bound beast to elegant angel.

So that's the mechanics of getting a Spitfire into the air, but what does it feel like to fly one of these beautiful machines 3 km above the clouds? How did it feel to be one of the women of the Air Transport Auxiliary or an American RAF pilot from the Eagle Squadrons, flying the one plane they'd all yearned to take to the air?

Once he's safely completed his mission for the day, Mark takes me back to his very first Spitfire flight. He was, he admits, scared. There's no such thing as a professional Spitfire flight simulator and the Battle of Britain Memorial Flight doesn't own one of the rare twin-seat Spitfires, so Mark's first flight was a solo one, in a Mk IX Spitfire that flew over the Normandy beaches on D-Day. He didn't get a wink of sleep the night before.

Once airborne, though, all his worries immediately dissolved in the clouds. He felt he'd left his brain on the ground and could simply enjoy himself. As he returns to that day, Mark's superlatives flow – she was sublime, so responsive, so quick, almost viceless. The controls, he says, are light and beautifully harmonised. She has the most benign stalling characteristics, so that even if you mishandle her, you're given plenty of warning. Rather than strapping yourself *into* the aircraft, it feels as though you've strapped it *on*:

> It becomes part of you. It's almost like the Spitfire knows what you want it to do. It takes the lightest of touch to respond to every input you make. It does feel

almost like you've grown wings. She does everything the moment you ask for it, if not before.

Landing, though, was as much of a challenge as he'd been warned. It's the one aspect of flying the Spitfire that has actually got harder than it was for pilots in the 1940s. They generally landed on grass fields, so could choose and alter their angle of approach in response to prevailing crosswinds. BBMF pilots land on a tarmac runway with a limited angle of approach. At the end of Mark's ecstatic first flight, he successfully landed with barely a bump. His relief, though, was short-lived. The control tower reminded him that he had to take-off and land twice more to satisfy his training instructor.

Mark has enormous respect for his wartime predecessors. In Battle of Britain combat, they could fly straight and level for just twenty seconds. Any longer and they were easy meat for a Messerschmitt. He knows the pressure that combat pilots were forced to put on their aircraft and on their own bodies. In the twists and turns of dogfights, they would pull and push their aircraft to a point where they risked unconsciousness.

High-speed dives place the body under positive g-force, or gravitational force equivalent. Under high levels of positive 'g', blood is moved from the head toward the feet. With a pilot's brain starved of the oxygen carried by their blood, they could lose their vision briefly or black out entirely. To Mark's relief, these seventy-five to eighty-year-old Spitfires are never put under that kind of pressure, and when he's flying combat missions in modern aircraft, he wears a g-suit, which inflates around the pilot's lower extremities to prevent blood from pooling in the feet and legs, maintaining the blood flow close to the brain.

Second World War fighter pilots were sometimes wary of talking about their feelings from those dark days, but, chatting to a fellow Spitfire pilot like Mark, many opened up their hearts. Mark has been left in no doubt about the day-to-day fear that went with their job. In his Eurofighter Typhoon, the modern weapons systems provide what Mark calls 'stand off'. He can engage targets long before they can see him or he can see them. Back then, says Mark, it was eye to eye, toe to toe – a visceral experience of man versus man, machine versus machine. You don't have to dig too far beneath the humour and bravado in the letters and accounts of wartime fighter pilots like Geoffrey Wellum, Art Donahue and Jack Lawson to discover that undercurrent of primal fear.

Some of Mark's best days in the Royal Air Force have been those when he's flown a Eurofighter Typhoon in the morning and a Spitfire in the afternoon. He warns me that you have to make a very conscious mental break between the two aircraft. Each might be the RAF's premier fighter of its day, but there the similarities end. Combat pilots rely on instinctive motor skills to allow for a greater focus on the variabilities of each mission. It's vital that you're not relying on the motor skills applicable to the Typhoon when you switch to the Spitfire. The Eurofighter Typhoon is operated under a 'fly by wire' system – there is no direct mechanical connection between the pilot's movements and the control surfaces of the aircraft. Computers interpret the pilot's inputs, adjust for external atmospheric conditions and correct mistakes. The computer adjusts flaps and rudders constantly to maintain the required speed, direction and stability. Shut down the computers and the Typhoon is impossible to fly.

Veterans have told Mark how the Spitfire was a quantum leap from the aircraft in which they'd had their initial flight training. It was so fast that they struggled mentally to keep up. In the Typhoon the computers handle all of the basic flying in order to leave the pilot free to concentrate on the navigation and weapons systems. The Spitfire and the other single-seat, high-speed fighters of the Second World War arguably provided the greatest mental challenge any pilots have ever had to master. That intense pressure is central to the stories of the young men and women who took on these extraordinary planes – often with the bare minimum of training – and helped win a war with them.

Few of those veterans have talked as vividly about the transcendent highs of flying a Spitfire, or written as honestly about the deep, dark lows of fighting and killing, day after day, as Geoffrey Wellum:

> You had to realise very early on in the Battle of Britain, if you stayed still, if you flew straight and level for twenty seconds you were dead. Never stay still. If you can't see anything it was always the chap you couldn't see who shot you down.[2]

Just seventeen when he joined the Royal Air Force, Geoffrey Wellum was the youngest Battle of Britain Spitfire pilot to survive the war. He credited his survival to luck – and to that twenty-second rule. But this level of tension, this hyper-alert state demands a toll:

I realised it at the time and I can confirm it now that I'd reached the pinnacle of my life and nothing would replace being in a fighter squadron for 1940, 1941 and 1942. And it never did.[3]

Geoffrey Wellum was brought up in Essex, where his father was a member of a local flying club. Young Geoff was immediately taken with the colourful biplanes and decided that there was absolutely nothing he wanted to do with his life but fly. In early 1939 he was a boarder at Forest School in north-east London. He was the popular captain of the First XI cricket team, but Geoff was eager to move on, desperate to take to the air:

I am seventeen and a half years old and, I suspect, a rather precocious young man. It was some six months ago when I first wrote to the Air Ministry. I was leaving school within a year and very much wanted to fly an aeroplane, so could they give me a job, please? It must have been a frightening prospect because they certainly took their time replying, but eventually I received a response in the guise of an enormous and rather complicated form together with a covering letter.[4]

It was August 1939 when Geoff was accepted for training with the Royal Air Force. He flew his first solo flight in a Tiger Moth biplane two days before Britain declared war with Germany. Training was fast and ruthless – navigation, instrument flying, aerobatics – too much to take in. Geoff and his fellow trainees were pushed to the limit. His first close RAF friend died when his aircraft stalled and crumpled to the ground. Others were sent home – given a 'bowler hat', in RAF slang – as they failed their flying tests or found themselves labelled as 'temperamentally unsuitable for combat'.

As his flying hours mounted up, moving from basic training on the Tiger Moth to advanced training on the Harvard monoplane, Hitler's forces flattened Poland, snatched Denmark and Norway, and bulldozed their way through Holland, Belgium and France to the English Channel.

In June 1940, with no combat training, having never flown a fighter aircraft, Geoffrey Wellum was assigned to a newly formed Spitfire squadron at Duxford near Cambridge. As his new comrades of 92 Squadron flew off to protect the British and French troops trapped on the beaches of Dunkirk, Geoff got his first taste of a Spitfire.

Spitfire K for King stood waiting. With its narrow legs, it seemed to Geoff like a delicate racehorse rather than a war machine. The dreamy teenager was grabbed by rough, experienced hands:

> The ground crew strapped me in and it was all a bit intimidating. Even the start-up, smoke coming back. I remember taxiing out being very careful. It seemed to hurtle itself into the air with me hanging on to the stick and the throttle, dragging me along with it.[5]

Once airborne, Geoff composed himself and gained control of the powerful machine:

> Everybody spoke about Spitfires and here I was flying one and I wasn't yet nineteen years old. There you

are careering around at 300 knots in this thing. Just the slightest thought that goes through your brain, conveys it to your hands and feet and the next thing you know, the aeroplane's doing it. You feel detached from fellow man.[6]

The exhilaration, mirrored by a strange, ethereal sense of calm, was like nothing Geoff had experienced before. There's really no other word for it: this was love.

It's clean, it's pure, it gets into the very soul. It gets into your soul and you think this is absolutely beautiful.[7]

After Dunkirk, the intensity of the battle eased. Hitler paused to plot his next move and Geoff got a chance to get to grips with his Spitfire. Thirty hours of flying this beautiful machine. It's barely enough, but it's more than most will get when the war reignites. Combat training was short, sharp and ineffective, so Geoff filled in the gaps by talking to more experienced pilots, developing his own philosophy of life and death.

Obviously we were going to be involved in a pretty serious business. Being shot down didn't appeal to me so I thought how do I avoid it? Make yourself a difficult target. How do you do that? Never fly straight and level.[8]

Fifteen seconds of ammunition, eight guns, three hundred rounds per gun. That's all Geoff would have to bring down the enemy and get himself out of trouble. Every burst would have to count.

Geoff was desperate to see some action, but now based in south Wales, 92 Squadron was a long way from the

French coast. During the day they chased the occasional lone raider in the Bristol Channel and, as the sun set, Geoff's training continued with a crash course in night flying. The Spitfire was never designed as a night fighter and Geoff struggled with his spatial awareness in the crow-black dark of the rural Welsh night.

Coming in to land one evening, the flares marking the runway appeared to vanish. He regained some height and tried again. Once more the flares vanished. On the third attempt Geoff suddenly realised that he was flying much closer to the ground than he'd thought. Fear surged, then ebbed, replaced by a brief moment of calm as he reconciled himself to death. A blinding flash of white light and the screech of tearing metal. It was a few seconds before he realised that the plane had stopped and he was still alive. The engine was off but he had no memory of killing it.

Geoff had crashed into a landing light and written off his Spitfire. He's grounded by his furious Commanding Officer. With no flying and little else to do, Geoff fell into a depression.

By now, though, the Battle of Britain was underway. Hitler was set on a summer invasion, but to cross the English Channel, he needed control of the air space over southern England. German bombers began targeted attacks on ports, airfields and aircraft factories. 92 Squadron was transferred to Biggin Hill in Kent, right on the front line. Geoff had made mistakes, but his CO could see that he was a decent pilot and every one of those was needed for combat duty:

We were posted up to Biggin on September 7th when the battle was at its height and replaced 87 Squadron that had been shot up and knocked about a bit and the first time I ever went into real conflict was there and I'd had thirty hours in a Spit.[9]

From the moment they arrived at Biggin Hill, 92 Squadron knew they were in the heart of the fire. On the ground, prey to surprise attacks, and in the air, up against appalling odds:

We were in a vast panorama of blue sky with the green contrasting fields of England.[10] We were vectored on to 150 plus coming in over Dungeness and I saw this mass of aeroplanes – looked like a lot of gnats on a summer's evening – and I thought, these chaps mean it. This is serious. That's the first reaction I really had. There's that dreadful thing – where do we start on this lot?[11]

It was the start of a fortnight of non-stop fighting that would define the rest of Geoff's life. Four or five sorties every day. He was already aware of the importance of this moment, not just for the country, but for his soul. The intensity of the conflict made him hypersensitive to everything around him. Every moment, every detail imprinted itself on Geoff's memory, every detail from dawn 'til dusk:

I used to walk out to my aeroplane first light. There were the ghostly images of the Spitfires and people working on them. You'd walk out with your parachute on your shoulder across the grass. You'd notice things like dew on the grass. It had been a lovely night, going to be a gorgeous day. And you'd probably sling up a little prayer or something. You wouldn't tell anybody else about it.[12]

Engines start. Merlins – ten, twelve, maybe fifteen of them. Their power restrained, forcibly held back until it seems as though the sky itself is urging them forward. The Spitfires judder and twitch. The grass flattens behind them.

Radar picks up a formation of enemy aircraft approaching the Kent coast. Plotters guide ten Spitfires of 92 Squadron toward one hundred or more German aircraft – they're bombers protected by Messerschmitt Bf109s – every bit as fast and deadly as the Spitfire.

Geoff catches sight of a Heinkel 111 bomber isolated from the protection of his formation. He gives chase and fires his eight Browning machine guns until he's out of ammunition and the smouldering bomber spirals toward the waves.

There's a moment of surprise, a brief surge of joy. It lasts just a few seconds – ten perhaps, maybe twenty.

Suddenly there's a flash and something flaming red streaks past the cockpit:

> More explosions behind me; a frightening noise I can't place as more tracers go tearing past. It is only a fraction of a second before it registers, but even that delay is too long.[13]

He's been hit, bounced from behind. Geoff snatches a look over his shoulder. There's a Messerschmitt tight to his tail. It turns with him and there's a staccato bass burst of thuds in the fuselage behind.

The Spitfire turns and turns tighter. Geoff starts to black out. The g-force is simply too much. Too much for Geoff, but maybe too much for the German. The Spitfire gains a few seconds and Geoff comes to with a rush of adrenaline:

> When that 109 caught me after that Heinkel, I remember thinking, if you get away with this, you get away with anything. Because he should have killed me.[14]

Geoff scans the sky above and below. He can't see the Messerschmitt, but that doesn't mean it's gone. He takes the Spitfire down to ground level, right down below the tops of the highest trees. It makes him an awkward target. Villages flash past, church towers, a startled woman looks up from her washing. Back up to 300 feet, a glance around and the Messerschmitt is gone.

Geoff feels an uncanny sense of serenity as he nurses his damaged Spitfire home:

> It happens particularly at the end of a day flying back on your own, you get the overwhelming impression you've survived and you get a peace as if there's an unknown presence with you. You're not alone – you can't explain it. Fighter pilots know it. You just know it's beautiful and after that peace you get a sense of loneliness. You've survived, it's getting dark. I want to get down to my fellow men, my fighter pilot mates. Unless you've experienced it you don't understand. Only fighter pilots understand it.[15]

The peace, the sensation of being in the hands of a greater power, lasts only as long as the flight home:

If you got to 5 o'clock you thought, 'the day though gavest us, Lord, has ended'. Straight off to the White Hart at Brasted, perhaps a game of darts, suppressing thoughts of mates that hadn't turned up and generally knocking back the pints. If there was pretty jewellery around you tried your luck.[16]

Every day, another comrade was gone – Butch, Bill, John, Crawford. The suppression of emotion was essential – if you lost focus, you would join the dead. There was no time to mourn. Not yet:

> We were young. You accepted it. It didn't register with you, not until very much later. Poor old 'so and so' hasn't come back today. Never mind, we're off to the White Hart now. If you didn't put it behind you, if you let your imagination run away with you, you were finished.[17]

Fear and boredom. First one, then the other. Day after day, the same routine. Take off at first light, engage and kill the enemy. Back to base to refuel and rearm. Mugs of tea as you wait, exhausted, to see how many of your comrades make it back alive. The stress builds and intensifies:

> We fought at first light, high noon, evening, dusk. It was a relentless ritual and we had no idea when it would end. All that mattered was getting airborne, fighting the war, surviving, until the next day. Day after day after day.[18]

Geoff still had his moments of teenage bravado:

> The waiting was the problem with me. I hated it. But the moment I got in that aeroplane and felt the vibra-

tion of the engine through the seat of my pants and I was strapped in and the ground crew got off the wing and waved, I felt, okay it's up to me.[19]

But looking back, there was nothing 'okay' about Geoff's mental health.

We all had combat fatigue in the end. Didn't realise it but we did. Exhaustion came into it and if you got too tired that's when you lost your concentration and (thought), 'eh, I can't be bothered, I can't be bothered, come on, kill me.'[20]

By the end of the Battle of Britain, nearly 2,000 German aircraft had been destroyed by the Royal Air Force, but the toll had been horrendous. So many dead – and so many survivors damaged for life.

Geoff and his Spitfire were sent to the island of Malta to help protect this strategic forward base for the Allies in the Mediterranean. His twenty-first birthday was spent on a Royal Navy aircraft carrier, HMS *Furious*, passing through the Straits of Gibraltar into one of the most dangerous regions on earth for a fighter pilot.

By the time Geoff arrived, Malta had been under Italian and German siege for nearly two years. Food was running out and fuel and ammunition were strictly rationed. The island's Hawker Hurricanes had been decimated by attacking Messerschmitt 109s and the latest Spitfires were desperately needed to repel the relentless bombing campaign. Geoff's first job was to fly one of thirty-eight Spitfires 650 miles (1,000 km) from the aircraft carrier to Malta, refuel and turn back to protect the ships he'd just left – a vital convoy bringing fresh supplies to the battered island. He was aware of a painful throbbing behind his eyes as he flew back across the Mediterranean. The area where the convoy should have been was empty, but he eventually spotted a battle cruiser and two merchant ships. It was clear that the close-knit convoy had been broken up and many ships lost. Geoff chased off some Junkers 87 bombers and returned to base.

Geoff spent the next two days patrolling the Mediterranean supply route before the remains of the rescue fleet finally limped into Valetta's Grand Harbour. Crucially, the oil tanker *Ohio* was among them, lashed to two naval destroyers. She'd been hit by bombs and torpedoes, her engines were out of action and there were fires burning on her deck, but the precious cargo of fuel was intact. Nine of the fourteen merchant ships had been sunk, as had the aircraft carrier HMS *Eagle*, one cruiser and three destroyers but the arrival of the Spitfires and the fuel eased the pressure on the island.

Geoff's headaches got worse. Malta was one mission too many for the exhausted fighter ace:

> I had a breakdown. I think they call it combat fatigue at the moment, but I'd been operational flying for three years and I had a very acute sinus which was very painful and I had quite a nasty operation there and everything went 'wuh'. I just collapsed.[21]

Back home, fishing in the ponds of Epping Forest, Geoff was initially relieved to be out of the cockpit but his anxiety soon returned. He felt he was no longer pulling his weight for the war effort and he was deeply disturbed by the distress in his mother's eyes when she looked at him: 'She was rather upset to see her son like this – that I wasn't the young man that went away – but it's called war.'[22]

Geoffrey Wellum, a hero, was broken:

> Something inside me gave way and I stretched out on the same familiar comfy bed and broke down. I grieved for my lost friends and I cursed that I had reached the pinnacle of my life before the age of twenty-two. Two full tours on Spitfires, the Battle of Britain and nearly 100 escorts and fighter sweeps over occupied France and I felt destroyed by the war.[23]

Nothing in his life ever compared with that desperate, deadly summer as a teenage Spitfire pilot.

18

Women Take the Controls

Just across the road from the southern edge of Hyde Park, there's a distinctive curved mansion block. It forms the junction of Knightsbridge and the Brompton Road. It's known as Park Mansions, with the Scotch House clothing store on the ground floors and comfortable apartments for the well-to-do above. In September 1945, a few days after VJ Day, a Supermarine Seafire roars past Harrods, tips its wings in salute, turns the corner at the Mansions and speeds west, bound for the naval air base at Culdrose in Cornwall.

At the controls is twenty-four-year-old Monique Agazarian. It's her very last day as an ATA girl and she's determined to show her mother in Park Mansions just what she's been up to for the last two years. It's also a tribute to her much loved brothers – Noel the fighter ace, shot down over North Africa in 1941, and Jack, tortured and hanged as a spy by the Gestapo.

Monique Agazarian – known to her friends as Aggie – was described in wartime newspaper profiles as a petite and vivacious beauty, with striking dark hair and eyes. One of six children of a French mother and an Armenian refugee father, Monique had longed to fly since seeing a production of Peter Pan in the West End, repeatedly leaping on to her bed from the top of her wardrobe. Her mother spent £5 on a derelict Sopwith Pup biplane as a playground toy for their Surrey garden. Aggie was hooked for life.

Nineteen when war broke out, she had just left her Paris finishing school and was studying to be an actress at the Royal Academy of Dramatic Arts. She was bound to be pressed into war work, so Aggie seized the initiative and joined the Voluntary Aid Detachment, working as a nurse at the Queen Victoria Hospital, East Grinstead. There she assisted the pioneering plastic surgeon Archibald McIndoe, treating the deep burns and serious facial disfigurements suffered by so many pilots in the Battle of Britain. The ugly reality of aerial combat did nothing to blunt her enthusiasm. Aggie was determined to fly and, for a young woman in 1940, there was just one possible route to the cockpit – the Air Transport Auxiliary.

The ATA was founded in 1939 as the logical answer to a logistical problem. By 1945 it had become a beacon of gender equality and given a generation of remarkable young women the chance to fly a Spitfire.

In the final months before the outbreak of war, Britain desperately needed more fighter planes to combat the threat from Germany's bomber squadrons. Britain's factories ramped up production as quickly as they could, but there was a crucial gap in the supply chain. How were these new planes to reach the bases of the Royal Air Force? If they're damaged, how would they get to the workshops for repair? The obvious answer was to fly them.

But what if all your pilots are busy fighting a war? What then?

The answer was to call on civilians with flying qualifi-cations – men too old for the RAF, men with injuries or disabilities that ruled them out of combat and maybe, just maybe, women.

The First World War had utterly transformed aircraft technology, and in the peace that followed, an exclusive new leisure industry was born. For a particular strand of the British upper-classes, to have your own plane was absolutely the thing.

'It was tremendous fun. Everybody went charging off, more or less anywhere.' Joan Hughes, twenty-one when war broke out, had been flying solo in her own plane since the age of 15:

> It's a sort of disease, it's a bug, and you know once you've got it you can never seem to get rid of it. Just to sort of be in the air . . . one really can't explain it. If I was suddenly stopped flying I don't know what I'd do. I would rather have flown than eaten really.[1]

Popping across to France for a day at the races or the casino, airstrip hopping over the German Alps or flying low across the Great Hungarian Plain. For those with time, money and a plane, the world really was your oyster.

Rosemary Rees, daughter of a baronet, and the future Lady du Cros, spent her winters skiing and her summers flying all over Europe:

> It was really the golden age of the private flyers; it's never been so much fun since. There was no wireless, nobody yelling at you, no commercial aircraft saying you got in their way. You just went. Every Aeroclub used to give rallies and parties over the summer. Oh, it was very great fun.[2]

One member of this flying elite was a wealthy banker, the son of a Parisian baron. He was called Gerard D'Erlanger, but everyone knew him by his nickname, acquired at Eton – Pop. As a director of one of the largest of the pioneering airlines, British Airways Ltd, Pop realised that his business would be shut down as soon as war was declared. Wouldn't it make sense for the government to put his planes and his pilots who were too old for combat to good use? The Air Ministry agreed and the Air Transport Auxiliary was founded in September 1939 at White Waltham in Berkshire, with Pop installed as Commandant. In the first days of war the ATA concentrated on freighting mail and medical supplies, but within weeks its volunteers began to replace combat pilots in the transfer of aircraft from factories and stores to front-line squadrons.

At this stage, the ATA was a man's organisation – D'Erlanger joked that it should stand for 'Ancient and Tattered Airmen'. No woman had ever flown an RAF plane. But these developments were being watched keenly by Pauline Gower. Daughter of a prominent Conservative Member of Parliament, she was one of those privileged few who had learned to fly between the wars. But unlike many others, she didn't fly just for fun. Pauline Gower was a shrewd businesswoman who had made her living giving joyrides and performing aerobatics for adoring crowds

By 1939, although still in her twenties, she was an experienced aviator, and well respected in the aeronautics world. At eighteen, Pauline had been presented at court and spent a season as a debutante on the London social scene. She had toyed with music, photography and horse-riding, but eventually settled on the idea of carving out some kind of career in aviation. Her parents were horrified, but she raised the money for flying lessons by teaching the violin. After just seven hours of tuition, she made her first solo flight. She quickly earned her private pilot's licence and went on to become just the third woman in the world to gain a commercial aviation licence.

With her friend Dorothy Spicer acting as her engineer, Pauline set up a business offering joyrides and air taxi services from an airstrip at Wallingford near Oxford. The pair then worked for several aerial circuses, which toured the fairgrounds and village fêtes of a nation in the grip of an aviation mania. When her mother died, Pauline decided it was time to settle down and set about writing a book on women aviators with her friend and fellow pioneer, Amy Johnson. The outbreak of war roused her from her very brief retirement.

Pauline was a master of the art of gentle but unrelenting pressure and, with her society connections, she knew just where to apply it. She worked fast, writing to D'Erlanger and lunching with the Director General of Civil Aviation. She pulled strings, and called in favours.

The result was a commission to head up a brand new Women's Section of the ATA. Young Joan Hughes was among those Pauline had her eye on: 'Well we all had to go down to Bristol and we were flight tested, and then there was an agonising wait, until you were told if you were selected or not.'[3]

Joan was in luck – she was one of the chosen few – a group of women who became known as the First Eight. These women were all effervescent young aviation pioneers, but they were certainly not a diverse cross-section of British society.

There was Rosemary Rees, a dancer in musicals and revues before she was an airwoman, and Marion Wilberforce, daughter of a Scottish Laird, who was rumoured to transport prize bullocks in the back of her plane. There was Mona Friedlander, a Mayfair socialite who played ice hockey for Great Britain, and Winnie Crossley, who had been a stunt pilot for a flying circus.

These were brave, adventurous society women with a taste for adventure.

The First Eight needed a minimum of 600 hours in the air, far more than the hours required for the ATA men. Four of the eight were already flying instructors; the remainder had experience of aerial displays, long distance flying or army co-operation. Gower was acutely aware that their gender would mark them out for attention, so she attempted to seize control of their media presentation. The First Eight were unveiled to the press in January 1940. Joan Hughes, with her wide eyes and unruly brown curly hair tucked in a smart navy-blue side cap, appears rather nervous in those early photographs. She had good reason to be wary:

> Well, there was an awful lot of adverse publicity to begin with. Of course a lot of men said, 'what are these females doing, they're doing us out of jobs'. And we KNEW, you know, if anything went a bit wrong of course it would be screaming headlines, 'Women Smashing our RAF Aeroplanes, what are these women doing being allowed to fly them?'[4]

C. G. Grey, Editor of *The Aeroplane* magazine – incidentally, a noted Fascist sympathiser – was not at all amused by these confident young women:

The menace is the woman who thinks they ought to be flying a high-speed bomber, when she really has not the intelligence to scrub the floor of the hospital properly or who wants to nose around as an air raid warden, and yet can't cook her husband's dinner.[5]

Independent women who defied convention – and gravity – inevitably attracted salacious attention. Joan Hughes rather relished it:

Well they just thought you know that if you flew, you were obviously rather a sort of racy type, you probably smoked and drank and heaven knows! Not quite the thing really. We were regarded as a sort of joke. You know there were only eight of us, and they used to call us 'Atta-Girls' – A-T-A you know – 'here come the girls' and everyone used to roar with laughter. It was fun actually. Much nicer than people looking down their noses.[6]

Even the enemy noticed these ATA Girls. The Irish Nazi propagandist, Lord Haw-Haw, broadcasting from Germany, condemned them as 'unnatural and decadent women'. At first the British press treated them with suspicion, but editors soon learned that their readers were fascinated by stories – and photographs – of fragrant young women taking to the skies.

A rather sober 1942 propaganda film from the Crown Picture Unit dramatised the role of the Ferry Pilots for a cinema audience, presenting the pilots as 'vital links in the air chain that stretches from the factory to the skies above Berlin'. The director, Pat Jackson, was careful not to over-emphasise the role of attractive young women, but couldn't resist opening with the tag line:

A Spitfire screams into the distance and we picture a fighter pilot heading for a prowling Nazi. Possibly that pilot was a girl![7]

None of those 'girls' were likely to be heading for a prowling Nazi – their guns were never loaded – but it was certainly a dangerous, demanding job and the scrutiny of the ATA women added extra pressure that their older male colleagues never had to endure. As Joan Hughes explained: 'You DAREN'T have an accident; you were TERRIFIED of having an accident, because it would just muck it up for everybody.'[8]

Rosemary Rees agreed that any misstep would be a judgement not just on them, but on the flying ability of their whole section, indeed, their entire sex:

> We had the appalling responsibility, weighing on our shoulders. If a man took a thing up and broke it, it was just too bad, people do after all sometimes break aeroplanes, but if we'd broken one, immediately they'd say, 'you see, we said they couldn't do it'.[9]

At first, these pioneering women, all experienced pilots with hundreds of flying hours under their belts, were limited to flying easily replaceable aircraft – biplanes like the de Havilland Gypsy Moth – that the RAF used to train new pilots. All through the long, bitter winter of 1939–40, Joan flew these open-cockpit planes, for hours at a time:

> My God it was cold! Open aeroplanes – no heating, no nothing! And it didn't seem to matter what you wore, balaclavas, scarves the lot, you always seemed to get a draught going straight down your neck. It was

quite funny sometimes when you'd stop to refuel, your face was so frozen you could hardly talk. And you couldn't smile because your lips were frozen.[10]

While the First Eight ferried training aircraft around the UK, the Battle of Britain was raging. The lives of 544 pilots and crew from Fighter Command were lost. Another 1,000 from Bomber and Coastal Commands also died. Britain's survival hung in the balance. The RAF needed more: more men coming through the pilot training programmes, more planes coming out of the factories and more delivery pilots to connect them.

Pauline Gower spotted an opportunity to expand her empire. She wrote to ATA Commandant Gerard D'Erlanger, imploring him to give her girls a chance to help. There was no reply, so, in typical fashion, she navigated around the obstacle and invited Air Marshall Christopher Courtney to lunch. As a member of the government Air Council with responsibility for RAF supplies, he was an influential figure. Impressed by her case, he took her request right to the top.[11]

Pauline soon got the green light she'd been waiting for. She could finally expand the women's section of the ATA, take on more pilots and teach them to fly larger and faster aircraft. By the middle of 1941, with two years of duty and an enviable safety record behind them, Pauline was instructed to bring five of her best women to be flight tested on something with a little more bite. It was a pivotal moment. Joan Hughes was one of the chosen five:

They then said, 'all right, we'll let you have a go at the Hurricane'. And five of us were allowed to do a

circuit in a Hurricane! Of course one was a bit nervous that one might make a mess of it, but one was
so terribly pleased to be doing it.[12]

The first of the five to slip into the Hurricane's cockpit, the first woman to fly a modern RAF fighter, was
Winnie Crossley. She was a well-known figure in inter-
war aviation, flying with C. W. A. Scott's aerial circus and
towing advertising banners over London. The high society
magazine *Tatler* photographed Winnie and her twin sister
Daphne in 1935, describing them both as 'experts in
the art of Aerobatics'. They may both have been fliers,
but the two tall, elegant girls were dressed in contrasting
fashion; Daphne in heels and fur coat, Winnie in a leather
and fleece flying jacket. Even then she looked ready to test
her skills on something more powerful than the sisters'
Gypsy Moth.

Sure enough, Winnie took off, looped the Hurricane,
landed smoothly and taxied toward her friends on the
apron. 'It's lovely, darlings,' she declared, 'a beautiful little
aeroplane'.[13]

Joan was equally delighted with their promotion to the
world of high-speed aviation:

You know it was an enormous thrill, it really was. And
having done that we were then allowed to fly Spitfires
and then . . . it was away.[14]

From that moment on, the most experienced women
of the Air Transport Auxiliary were free to fly any of the
RAF's fighters or twin-engined bombers. By 1943, Joan
Hughes and Lettice Curtis were cleared to fly some of the

biggest planes of the war – four-engined bombers such as the Shorts Stirling and the Avro Lancaster.

As the war progressed, the ATA grew and grew – the men's section, and the women's. By the middle of 1942, they were recruiting pilots from all over the world. Women, sensing an opportunity they couldn't dream of in their own countries, applied in droves. Airwomen from the USA, from Chile, Australia and South Africa made the perilous journey to the UK, to sign up and serve with the Air Transport Auxiliary.

With her pilots in such demand, Pauline Gower was cleared to take on new recruits *ab initio* – women with no flying experience who would be trained from scratch. That was terrific news for a dark-haired ball of laughter, Monique 'Aggie' Agazarian.

Ever since hearing about the ATA, Aggie had bombarded the Auxiliary management with letters pleading to join. Each letter told a different story. Sometimes she admitted that she'd never flown, sometimes she claimed all sorts of experience in all sorts of aircraft. When she finally reached the second stage of recruitment, her interviewer commented that 'you seem to have an extraordinarily varied flying career.'

Aggie somehow convinced the interviewer of her aptitude, but immediately failed the next test – at 5´3˝, she was too short to qualify. It was suggested that she try lying down on the floor before being measured again. Somehow that did the trick. The crucial extra inches were gained.

Aggie was one of ten women accepted for basic training, but her passage to the cockpit was far from smooth.

Her infectious energy and eccentric sense of humour seemed to suggest to her superiors that she wasn't taking flying seriously. An early training report cleared her to fly, but labelled her, rather portentously, as 'not a natural pilot'. Nothing could have been further from the truth:

> The commanding officer, when he took me for my wings test, not a smile out of him, a face like an old boot all the way through and instead of taking an hour and a half it took two hours! And I was so dejected by this time, I thought 'oh no'! Anyway, he got out and he turned around and said, 'you were very good!'.[15]

At first Aggie flew the 'taxi', ferrying more experienced pilots to their planes, but she soon spread her wings. In just six months Aggie would fly twenty different models of aircraft. She decorated her log book with a picture of each plane. By the end of the war there would be muscular Mustangs, elegant Rapides and clunky Fairchilds as well as Grumman Hellcats and Hawker Hurricanes. She would often fly several different planes in a single day.

As with all ATA pilots, she relied upon her 'bible' – the Ferry Pilot Notes: a loose-leaf folder with a page for each plane, giving the most basic of instructions, such as the take-off speed, the landing speed and the stalling speed.

There were so many planes to fly, and so little time to master them, but there was never any question which aircraft Aggie longed to take to the air.

It took the First Eight women two years to be allowed into the cockpit of a Spitfire. By 1943, it took Aggie a mere sixty hours of flying.

This meant so much to Monique. Her big brother, Noel, had flown a Spitfire right through the Battle of Britain, shooting down six German aircraft. One of his Spitfires, R6915, is still on display in London's Imperial War Museum. In 1941 he was transferred to 274 Squadron, based in Libya, where he was shot down and killed by a Messerschmitt 109. To experience the same highs as Noel brought him close to her once again:

> It was the most wonderful aeroplane in the world. You didn't have to fly it, you just thought about it. I used to waltz round those great big fluffy clouds. The Blue Danube mostly, you could waltz in and out, it was marvellous. It really was beautiful to fly.[16]

For so many of the ATA women, the Spitfire was the ultimate prize, an absolute treat to fly. American pilot Helen Richie called the Spitfire 'a fish through water, a sharp knife through butter, a bullet through the sky'. Joy Lofthouse immediately felt she'd come home:

> The Spitfire was the favourite. You felt part of it. The cockpit was very small, the wing span didn't feel very big. It was the nearest thing to being a bird and flying oneself that you knew. Very light on the controls, she turned with a slight touch on the stick.[17]

Jackie Sorour from South Africa was the ATA's youngest pilot when she joined up at the age of eighteen in July 1940. Her mother had wanted Jackie to study at Oxford University, but reluctantly agreed to fund an alternative Oxfordshire training course – at Witney Aeronautical College. Trapped in England by the outbreak of war, she joined the Women's Auxiliary Air Force, working as a radar operator in the early stages of the Battle of Britain before transferring to the ATA at the first opportunity. Jackie would never forget the impact of her first Spitfire flight – her emotional response was raw, sensual and immediate:

> It seemed the most natural thing in the world to be sitting in the cockpit, as though my entire life had led to this moment. I started up inexpertly and felt the power coursing through the Spitfire's frame. A little awed but stimulated by the urgent throb of the Merlin engine, that seemed to tremble with eagerness to be free in its own element, I taxied cautiously to the downwind end of the field, carefully I familiarised myself with the controls. As the ground fell away at fantastic speed, I felt exhilarated by the eager, sensitive response. Singing with joy and relief, I dived and climbed and spiralled, round the broken clouds.[18]

ATA pilots rarely had any radio communication with the ground. Once they took off they were completely on their own. For Joy Lofthouse, a twenty-year-old Lloyds Bank cashier from Gloucestershire, those hours, alone at the controls of a plane, as independent as any woman of the 1940s could ever be, were close to heaven:

> Once you had taken off it was almost like being a bird. You had no contact with the ground, no radio contact. Nobody could tell you that you were doing the wrong thing. It wasn't so great in bad weather but on a nice day especially on the south coast it was just divine, with the sea on one side and the countryside on the other.[19]

Weather was an understandable obsession for ATA pilots. Their bosses decided early on that training would not be given in flying by instrument. They wanted their pilots to avoid risk. If they couldn't see the way ahead then they shouldn't be flying.

Pilots were expected to stay low, avoid enemy contact and navigate by roads, rivers and railway lines. If cloud or fog appeared unexpectedly, they were under strict orders to land at the first opportunity and wait until the weather cleared.

Sometimes that time could be put to good use. Joy Lofthouse made a habit of breaking her journeys at American Air Force bases: 'I suppose a lot of us looked quite young and glamorous in our uniform and they hadn't seen too many women pilots before because we were the trailblazers.'[20]

Charming officers would lead her to the PX – the US equivalent of the NAAFI – where stockings, lipstick, chocolate and other hard-to-find luxuries could be bought.

*

Bad weather sometimes offered a welcome break from a tough flying schedule. More often it led to stress, scrapes and sometimes injury or death.

Amy Johnson made her name in 1930 as the first woman to fly solo from Britain to Australia. Piloting her flimsy Gypsy Moth, laden with extra fuel tanks, from Croydon Aerodrome, she braved desert storms over Iraq, a crash in India and the shark-infested Timor Sea. Twenty days later she arrived in Darwin to find herself a global star. Songs were written about her, celebrated couturiers begged her to wear their clothes, hairdressers were asked to recreate the 'Johnson wave' and a new variety of rose was named after her.[21]

Johnson was the most experienced woman pilot in the Air Transport Auxiliary, experienced enough to ignore the rules. On 5 January 1941, she took off in poor weather from Blackpool in an Airspeed Oxford, a twin-engined trainer aircraft. Bound for RAF Kidlington in Oxfordshire, she struggled to find a break in the cloud cover. With limited instrumentation and no radio communication, she became disorientated and found herself circling the Thames Estuary until she ran out of fuel and was forced to bail out. Her parachute opened and she landed alive in the water. She was spotted by a convoy of warships, but sea conditions were rough and snow was falling. The crew of HM Trawler *Haslemere* tried to reach her with ropes, but as the ship rocked and rolled in the swell, Johnson was lost beneath the ship. Lieutenant Commander Walter Fletcher, captain of the *Haslemere*, thought he saw a second body in the water. He dived in to attempt a rescue but succumbed to the intense cold and had to be rescued by lifeboat. He died of exposure in hospital a few days later. Amy Johnson's body was never recovered.[22]

*

A total of 173 ATA pilots died during the war, sixteen of whom were women. Flying high-powered machines, surrounded by gallons of highly flammable fuel, there was always a keen awareness of the proximity of death. Despite that, the ATA women seem to have shared a remarkable ability to remain calm in a crisis. Margaret Frost, a vicar's daughter from Sussex, was another ATA recruit lacking in height if not courage. She was accepted into the ATA only when she bought some high-heeled shoes which she stuffed and stacked with cardboard. She would fly most aircraft sitting on a parachute for a better view. The P-51 Mustang was the worst, she said, noting that 'Americans thought we were all six foot four'. Her seat once gave way when she was taking off in a Barracuda torpedo bomber. She found herself having to fly blind from a horizontal position:

> I'm sure we never panicked. We were too well trained and too young. Your heart was in your mouth but you never panicked.[23]

Margaret remained calm and resolute, even when faced with the inevitable deaths of friends and colleagues in the ATA and RAF: 'I think there were so many ghastly things happening in the war that your feelings were blunted unless it was somebody you knew extremely well.'[24]

Weather was the biggest killer, but Britain's air defences caused their own issues, as Joy Lofthouse explained to the BBC: 'Other than the weather, barrage balloons were another big problem.'

Positioned above every major city, these enormous hydrogen balloons of nearly twenty metres in length were held aloft by thick steel cables, sometimes rigged with explosives, to ensnare any pilot fool enough to fly through them:

> If an airfield you had to go into was within the bar-rage balloon area, you had an alleyway you went in, through the balloons, but they changed that every day, so that the Germans couldn't get wind of the way in.[25]

Lettice Curtis, tall and serious, was one of the most experienced and technically proficient flyers in the ATA, but even she wasn't immune to the danger:

> One day I took off without checking, because I thought I knew, and flew in in the usual way, and flew out again. There was a lot of low cloud that day. When I got home that evening, I was told I'd flown through the barrage. I hadn't seen anything, I'd come through and missed the lot. Even now my stomach turns over when I think of it.[26]

The dangers were real, but for many women fliers, such as Monique Agazarian, their years in the ATA were filled with the comforting pleasures of good company. Aggie flew with the all-woman 'Ferry Pool' at Hamble near Southampton. On days off the Pool girls would take the train up to London, dance at the Grosvenor House Hotel and sleep off their excesses, safe from German bombs, in the shelter of the London Underground. These women would be Aggie's friends for life:

> It really was a slice of the most wonderful time of one's life. Rather strange because it's over the background of the horror of the war. And it was awful, you looked every day to see who'd been killed, and you lived with it. I lost two brothers who I adored. But, the job really was the most wonderful thing in the world.[27]

The Air Transport Auxiliary was disbanded in November 1945. Peace had come, and its services were no longer required. But Pauline Gower, Senior Commander and

Director of Women Personnel for the ATA, recognised just how important their work had been:

> Had one weak link in personnel been allowed to mar the organisation, the Royal Air Force would have less aircraft today. A great and strong chain has been forged by the people of this country, and during the blood and tears and sweat through which we have passed, it has never broken. Lives have been given and are being given cheerfully, every day, and Ferry Pilots are among those who have made the great sacrifice.[28]

In the six years of the Second World War, the ATA ferried over 300,000 aircraft of 147 types. Male pilots outnumbered the women – 1,152 to 168 – but it was the women whose lives and opportunities were fundamentally transformed. It was the ATA girls who changed attitudes to women in aviation and the armed forces.

By 1943, the women of the ATA were not only doing the same work as the men, they were also being paid the same. Thirty years ahead of the Equal Pay Act, the Air Transport Auxiliary was one of the first significant employers to pay its male and female employees the same wage. Pauline Gower didn't want her pilots held up as superwomen – they were just as brave and skilled as their male comrades, no more and certainly no less:

> Some people believe women pilots to be a race apart, and born 'fully fledged'. Women are not born with wings, neither are men for that matter. Wings are won by hard work, just as proficiency is won in any profession.[29]

Despite the apparent progress, jobs in aviation proved hard to come by in 1945. For most of the ATA women, the end of the war meant the end of their flying career. Joy Lofthouse was heartbroken:

> When I finally had to leave ATA I felt devastated. I thought, what am I going to do with the rest of my life? I was so forlorn. One was losing touch with all one's friends and the life you knew for the last couple of years . . . Everyone felt like that. Really unsettled.[30]

Jackie Souror couldn't get the roar of the Spitfire out of her blood. After the war she joined the Women's Royal Air Force Volunteer reserve, flying Meteor and Vampire jets. When its training schools were closed down, she scouted around for more flying work, finding the perfect job ferrying ex-RAF Spitfires from Cyprus to Rangoon for the Burmese Air Force.

It was the future of aviation that interested Lettice Curtis. She worked as a technician at the government research facility, the Aeroplane and Armament Experimental Establishment, before moving to Fairey Aviation as the senior flight development engineer working on the Royal Navy's Gannet anti-submarine aircraft. But even she couldn't leave the Spitfire behind, competing in a series of races against the country's top test pilots, flying a Spitfire owned by the America air attaché to London.

Joan Hughes carved herself a satisfying and suitably glamourous niche, taking on stunt flying duties for moviemakers. Her tiny frame perfectly fitted the delicate replica of the 1907 Demoiselle monoplane in the 1965 comedy *Those Magnificent Men in Their Flying Machines*. She coached

Kenneth More for his role as Douglas Bader in *Reach for the Sky*, and finally got the chance to take part in a dogfight, flying replica First World War fighters for *The Blue Max*.

Perhaps her oddest screen role was as Lady Penelope's stunt double in *Thunderbirds 6*. As in the long-running sci-fi television series, the movie spin-off was populated by stringed puppets, but the director was determined to include a live-action flying sequence. Joan flew a Tiger Moth along the route of the unfinished M40 motorway with models of *Thunderbirds* characters clinging to its wings. To satisfy a watching health and safety official, she was supposed to touch down before she reached a bridge and pass beneath it on her wheels. According to Hughes, a crosswind caught the puppets, took her plane off the ground and forced her to fly under the bridge. She was arrested and charged with seven counts of dangerous flying. Hughes claimed, not entirely convincingly, that the incident had been the first time in her career that she'd been afraid. After a two-day hearing at Aylesbury Crown Court, she was acquitted of all charges.

Monique Agazarian found post-war life a little tougher, but she was equally determined to keep flying. She applied for work with airlines large and small, but found herself turned down again and again. Eventually she secured work with an air charter company, Island Air Services, flying flowers from the Scilly Isles and operating pleasure flights from Heathrow Airport. Within three years she was the Managing Director and Chief Pilot. Aggie went on to write the definitive manual on advanced instrument-flying procedures and she became one of the pioneers of flight simulators, training new pilots on her own full-movement simulator in Room 129 of the Grosvenor Hotel. Not at all bad for the woman dismissed by a male instructor in 1943 as 'not a natural pilot'.

Aggie died in 1993. At her wake an old ATA colleague played a cassette labelled 'Merlins for Monique'. As the distinctive roar of the Spitfire engine blasted from the speakers, hearing aids shrieked and the entire back row of elderly ATA veterans instinctively ducked before erupting in laughter.

19

The Yanks are Coming

'All Tiger aircraft, patrol Dover at 10,000 feet; patrol Dover at 10,000 feet.' For Art Donahue, this is his first flight with his new squadron. He's hoping for a first sight of France: 'Climb to 15,000 feet. Bandits approaching from the north.'

Then there's a barked 'Full throttle!' from the section leader, and Art sets the engine screaming with a surge of power. To the farm boy from America's Mid-West, it feels like the buck of a spurred horse. He flicks the guard off his firing button and switches on the gun sight. The French coast slides into view but there's no sign of German air-craft: 'Believe the enemy is now heading south and passing beneath you.'

For half an hour the squadron circles close to Calais until grey streaks are spotted to the east: 'Tally hoo!'

It's the Squadron Leader's order to pick your target.

Art tips into a screaming dive toward a German beneath him. The air-speed indicator needle eases past the 400 mph mark and the controls stiffen. Art's close to blacking out, but he levels his Spitfire and narrows his eyes as the mottled grey machine grows steadily larger in the circle of his gun sight. A short burst, 130 bullets, then Art's plane skips sideways as it hits the slipstream of the enemy. Over Cape Gris Nez now, the low green cliffs of France jutting out into the English Channel, the Messerschmitt turns to face him. But Art can turn tighter. His finger moves back to the firing button when there's a burst behind his ears, the plane shakes and there's a hailstone rattle, tat-tat-tat, on the fuselage.

Art has a second 109 right on his tail. For the next few minutes he turns and weaves, his compass spinning uselessly. He blacks out and comes to. He tries to shoot but his gunsight is dead. In the distance he spots a thin ribbon of white. The cliffs. Dover. He turns west at full speed and the Germans, out of fuel or ammunition, give up the chase.

> It was hard to realize that this had all actually happened and wasn't a dream. This was August 5th, scarcely six weeks from the time I had been at home in Minnesota, cultivating corn! That corn wouldn't be big enough to cut yet![1]

The newspapers in the United States lapped up Art's first British adventure, with the *Democrat and Chronicle* of Rochester, New York, quoting Art's breathless account of the battle: 'I was never so tickled in all my life – we went across the Channel to look for trouble, but it was the Germans who found it.'[2]

Art Donahue was one of eleven Americans to fight in the Battle of Britain, one of four who flew a Spitfire in action that summer. Art – short, strong and fair-haired – was born and raised on a dairy farm in St Charles, Minnesota. By nineteen he was the youngest commercial pilot in the state, making a precarious living through the

Depression in Texas and Wisconsin as a flying instructor, aircraft mechanic and occasional barnstormer – performing aerobatic stunts at State Fairs. He worked alongside Max Conrad, a pioneer aviator who went on to set numerous records for endurance flights around the world.

When war broke out in Europe, Art felt strongly that this should be America's war as much as that of Britain and France. Hitler he regarded as a dangerous menace. He applied for a commission in the United States Air Corps Reserve, but a series of frustrating delays left him no closer to flying combat aircraft. On the pilot grapevine he heard that the Royal Air Force was keen to hire Americans for non-combatant roles. Just a few weeks earlier, the United States government had issued a Presidential proclamation banning the recruitment of men for the armed forces of foreign countries. The proclamation wasn't just a symbolic sop to America's powerful anti-war movement – breaking the law could lead to the loss of US citizenship.

Young, fearless and desperate to fly, Art headed north, across the border to Ottawa in Canada. There was already a network of recruiters working there, funded in part by Charles Sweeney, a businessman and society sportsman. Born in Pennsylvania but based in Mayfair, Sweeney taught the Duke of Windsor to play golf and had married Britain's most eligible debutante, Margaret Whigham, the future Duchess of Argyll.

Margaret's life has its own place in the Spitfire story, linking several of its most important characters. At fifteen she was made pregnant by eighteen-year-old David Niven, who went on to co-star in the R. J. Mitchell bio-pic, *The First of the Few*. She later had relationships with Max

Aitken, the fighter pilot son of Lord Beaverbrook, and Duncan Sandys, son-in-law of Winston Churchill. Her second husband, the 11th Duke of Argyll, was divorced from Lord Beaverbrook's daughter. It's safe to say that the Anglo-American/Scottish-Canadian upper classes of the 1930s operated in a surprisingly small world.

In 1933 Margaret was engaged to the 7th Earl of War-wick when her head was turned by the dashing American golfer turned RAF recruitment officer.

Charles Sweeney recruited American citizens for the French Air Force and, after the fall of France, helped American pilots travel to the UK via Canada using thousands of dollars of his personal fortune. Art Donahue made contact with the Canadian recruiters and offered them his impressive log-book, showing 1,500 flying hours: 'In fifteen minutes I was out on the street again with a promise that I would be on the next boat to England.'[3]

Sure enough, within a fortnight, he was on-board the RMS *Duchess of Atholl*, a passenger liner converted into a troopship, sailing with Canadian soldiers across the dangerous waters of the North Atlantic.

Art wasn't the first American aviator to reach England with a view to joining the fight. Already flying were Eugene Tobin and Andrew Mamedoff, two friends from California, and Vernon 'Shorty' Keough from New York. At just 4′11″, Keough was the smallest pilot in the RAF, rumoured to need two cushions to see out of his Spitfire. Tobin came to Europe to help Finland fight its Winter War against the Soviet Union, but a peace treaty had been agreed by the time he arrived, so he joined Keough and Mamedoff in signing up for the French Air Force. As France fell to Hitler's *Blitzkrieg*, the three companions fled to England, where, as experienced combat pilots, they were warmly welcomed into the Royal Air Force.

But why were Americans willing to put themselves in such mortal danger in a war that wasn't yet theirs? The reasons were as varied as the colourful cast of characters who crossed the Atlantic.

Art Donahue was described by his 258 Squadron comrade John Campbell as 'an idealist – one of the few real ones in the squadron.'[4] Art himself wrote that: 'This was America's war as much as England's and France's because America was part of the world which Hitler and his minions were so plainly out to conquer.'[5]

Reade Tilley was a racing-car driver from Florida who joined the Royal Canadian Air Force in June 1940 and was commissioned in the RAF in August 1941. He went on to win the Distinguished Flying Cross as a Spitfire ace, with seven confirmed kills of German aircraft. Thirty-five years later he told the BBC:

> It was pretty apparent that help was needed here. We were interested not only in England but in flying and fighting and that is the primary reason we came over. What you want to be is where the action is, if you're inclined that way and I think most of us were.[6]

That was perhaps closer to the truth for the majority of Americans in the RAF. Their distaste for German aggression was certainly part of the equation, but, for many, the thirst for adventure was what really drove them to join up.

Colonel James Saxon Childers of the United States Army Air Force attempted to pin down their motivation in his 1943 book, *War Eagles*. Childers interviewed many of the American airmen who had served or were still serving with the RAF. Most failed to give him a straight answer – these young men were not the type to self-analyse, preferring to talk about British beer (warm), coffee (universally terrible) and WAAFs (just peachy). Colonel Childers, though, was a pretty good journalist for an airman, teasing out the true motivations beneath the banter and bluster:

> These young fellows left home for the reason that young fellows have always left home. There was for them, of course, the perennial excitement of what goes on beyond the horizon. There was, for these boys of their particular generation, the accentuated glamour of battle and heroes in battle. Loving airplanes as they do, they had read scores of books and magazines about the airplanes of the last war and the famous fighters who flew them . . . Here was really a chance for them to play the lead in 'Hell's Angels', to play opposite alluring platinum blondes like Jean Harlow that they had seen in the pictures and win them by incredible exploits in battle.[7]

In the recruitment queue, just behind the Mid-West barnstormers and crop-dusters, there was Billy Fiske, a Cambridge graduate and Olympic gold medallist from a New England banking family. Richard 'Indian Jim' Moore was the RAF's first – and possibly only – Native American recruit, while Jack Kennerly was abruptly thrown out of the RAF for his bad behaviour, on the ground and in the air. Narrowly avoiding an official court martial, Kennerly returned to the States, wrote a highly imaginative account of his brief wartime exploits, and was signed up by Warner Brothers as a technical advisor on *The Flight Patrol*, a war film starring Ronald Reagan.

Over 6,000 Americans are reported to have been recruited for the RAF or the Royal Canadian Air Force, but only eleven arrived and trained in time to take part in the Battle of Britain. Art Donahue was there right from the start, fast-tracked through training into 64 Squadron. After his first engagement with the enemy on 5 August, the action came thick and fast. On 12 August, Art was on the ground refuelling when news came in of a '450 plus' raid forming up over the French coast. Once airborne there was no sign of any large formations, but he later wrote of an encounter with a small group of Heinkel 113 fighters:

> They were good looking airplanes, and I remember that they were painted all white. We mêléed about a little, and I ended up getting chased down into the clouds.[8]

The Heinkel 113 didn't exist. It was an invention of Nazi Propaganda Minister, Joseph Goebbels. Pre-war prototype Heinkel 100s were photographed in a variety of colours and markings at air bases around Germany. The Heinkel 100 had been a rival to the Messerschmitt 109 in the competition to become the Luftwaffe's front-line single-seat fighter, but it never entered full production. The doctored photos were published widely in German newspapers and magazines. British Intelligence was taken in by the ruse and the word was passed to RAF fighter squadrons that there was a new – and possibly very dangerous – opponent in the skies. Whatever Art actually saw, he reported it being quick and smart enough to tuck in behind him, loose a burst of cannon fire and sever his elevator cables. Without those he had no control of the aircraft. Smoke was starting to fill his cockpit when another fighter fired a salvo:

> The din and confusion were awful inside the cockpit. I remember seeing some of the instrument panel breaking up and holes dotting the gas tank in front of me. Smoke trails of tracer bullets appeared right inside the cockpit. Bullets were going by between my legs. I remember being surprised that I wasn't scared any more.

Art's Spitfire was finished.

> A light glowed in the bottom of the fuselage, somewhere up in front. Then a little red tongue of flame licked out inquiringly from under the gas tank in front of my feet and curled up the side of it and became a hot little bonfire.[9]

Art tugged the locking pin and released his straps. The heat surged as he squeezed his shoulders out of the cockpit. The velocity of the plane's death dive dragged him out and scraped him painfully along the fuselage before releasing him into the open air. He pulled his parachute rip-cord, his body jerked and suddenly everything was calm, quiet and painless. The battle seemed miles away.

Art was badly burned. Rushed to a hospital near Canterbury, he could barely move without excruciating pain. He monitored the growing intensity of German bombing raids from the sounds he could pick up from his hospital bed:

> At first the sound would be like a distant storm approaching – just a heavy, distant murmuring and rumbling that gradually grew louder. It still sounded like a great wind approaching until finally distinct little individual sounds would separate themselves from the rest. The smooth high-pitched moan of a Messerschmitt in a power dive would rise above the rest of the sound momentarily, echoed by the sound of another doing likewise a few seconds later.[10]

When his burns finally healed, Art decided to use his remaining sick leave to see the sights of the capital. By day London was just as he'd imagined it – Buckingham Palace, double-decker buses and cheery cockney costermongers – but at night the city felt like a dangerous place to be. With dusk came the first of the sirens. Most of the residents of his lodging house rushed for the air-raid shelter, but Art decided to hang around to 'watch the show'. There were planes high overhead and anti-aircraft guns firing, but before too long the 'all-clear' sounded and Art returned to his room for a good night's sleep:

> About midnight I was suddenly awakened by a distant sound like steam escaping from a radiator. It was a ghostly sort of noise, like something slipping through the air in the distance at great speed, and it was rising in intensity. In perhaps four or five seconds it rose to a noise like that of a locomotive letting off steam close by, and then to a fiendish shriek, ending in a heavy explosion not far away, that shook the building.[11]

As Art tugged his pillow over his ears, the heedless courage of the Spitfire cockpit succumbed to an intense wave of stomach-churning fear:

> None (of the bombs) landed really close, but each sounded, before it hit, as if it were aimed for a point midway between the washbowl in one corner of my room and the suitcase under my bed. I had learned not to mind being shot at, very much, but I couldn't get used to bombing![12]

Next morning, a walk through the bomb-blasted City and East End seared dark and disturbing images on Art's consciousness. London was a mess, but for Art there was hope among the ruins. In one City of London street stood a dress shop. The walls outside were black from the fire; the windows cracked and the window frames scorched. Bricks and rubble blocked the street in front of it:

> But the sidewalk was swept clean, and that little shop, with ghastly fire-blackened desolation all around it, burned-out stone buildings towering around it on both sides of the narrow street – that little shop was still open for business and doing business, with its windows filled with a neat display of ladies' dainty underthings![13]

In 1941 Art took some leave back home in Minnesota. He used the Atlantic voyage to pull together his diaries and write his personal account of the Battle of Britain. It was published in the United States in August, one year on from his move to England.

Names of comrades and air force bases were changed by the British censors. There's little doubt that its publication was intended, whether by Art himself, by the publishers or by the British government, to add fuel to the pro-war cause in America. It was certainly a cause in need of a boost.

Joseph Kennedy, the US Ambassador to London and father of the future President, had loudly expressed his belief that Britain was finished. Some kind of accommodation with Hitler was not only unavoidable but advantageous to the United States. An even more prominent voice opposing intervention in the war in Europe was Charles Lindbergh, the world's most famous aviator. In 1927 he had piloted his custom-built plane, *The Spirit of St Louis*, on the first non-stop solo flight from America to Europe. As spokesman for the America First Committee, this national hero now divided his own country. His writings and speeches dripped with his intense hatred of President Roosevelt; they were laced with virulent anti-Semitism and they stridently opposed any form of support for the British war effort. The pro-war party needed all the help it could get from charismatic airmen like Art Donahue.

The Brits in Art's book are plucky in the face of adversity, the Germans wily and devious. British casualty figures should be trusted implicitly, while Goebbels and Göring massively under-estimate Luftwaffe losses. Despite this distinct whiff of propaganda, Art's descriptions of combat and his vivid pen portraits of London under fire ring with genuine feeling. Partnered alongside the iconic Blitz broadcasts of CBS Radio's Ed Morrow, Art's memoir gave Americans an emotionally honest account of how it felt to live in a nation under siege.

Art's words also captured the changes he felt in himself. In just one intense year, this softly spoken Mid-West farm boy had become a seasoned killer:

I switch on my gun sight and uncover my firing button and take it off safety. I do it absently, without the tremendous conflict of emotions that I had the first times last summer; and that is one of the very few evidences that I can see of any change that the war has made in me. I don't feel that it's hardened or toughened or aged me, but it does seem to have seasoned me to the point of nonchalance toward its savagery.[14]

While Art Donahue was enduring the savagery of air combat over England, a new wave of airmen was crossing the Atlantic. Twelve young men – six white, six black, all new recruits to the RAF – were photographed before they left their homes in Barbados. All of them sported broad smiles and sharp suits. The six black men, including future Spitfire pilot Arthur Weeks and the first Prime Minister of Barbados, Errol Barrow, were pioneers.

Black fliers had fought for Britain in the First World War, but a 'colour bar' had since been enacted, banning the recruitment of officers who weren't British-born, of British parents and of European descent. Desperation, however, bred a reluctant spirit of togetherness. Almost as soon as war was declared, the urgent need for manpower trumped the racist recruitment policy. The Army and Navy were slow to respond, but the Royal Air Force, as the junior service least resistant to change, began recruitment campaigns in the Caribbean and the colonies of West Africa.

At first glance it may appear strange that black residents of Jamaica and Trinidad would volunteer to fight for an Empire that had enslaved their great-grandparents and kept most in varying degrees of bondage and poverty since the abolition of slavery in 1834. Caribbean aircrew who left accounts of their time in the RAF, often made it clear that they were not fighting for the King or Empire but explicitly against the Nazi regime. Bomber Command navigator Sergeant Johnson, for example, said:

> Many people don't think about what would have happened in Jamaica if Hitler had defeated Britain, but we certainly would have returned to slavery.[15]

Around 500 black Caribbean men joined the RAF as aircrew, with another 6,000 joining as ground support staff. In a grim flashback to the days of the slave trade, some found themselves confined to the hold aboard American cargo ships on the slow and dangerous voyage across the Atlantic. Once safely in RAF hands, most men were treated fairly in training and in their allocation to squadrons. Skills appear to have been the deciding factor rather than colour. An Air Ministry Confidential Order of June 1944 stated:

> All ranks should clearly understand that there is no colour bar in the Royal Air Force . . . any instant of discrimination on grounds of colour by white officers or airmen or any attitude of hostility towards person-nel of non-European descent should be immediately and severely checked.[16]

That Order certainly implies that racist behaviour was being experienced by black officers and crew, but, at least in official terms, it put the RAF well in advance of the policies of most civilian employers in the United Kingdom prior to the 1970s.

In sharp contrast to the situation in the US military, there was no segregation in the Royal Air Force and black crew members mixed freely with their white British and European counterparts. According to the historian Mark Johnson, whose Jamaican great-uncle served in Bomber Command, black pilots were often regarded as mascots, providing squadrons with a distinctive identity. Many, however, were keenly aware that they walked a thin line between welcome novelty and perceived threat. Flight Lieutenant Billy Strachan, from Kingston, Jamaica, commented that:

> When you arrived anywhere as the first black man you were treated like a teddy bear, you were loved and fêted. Two they coped with; it was when three or more arrived that things got sharp.[17]

The only Spitfire squadron known to feature three black Caribbean pilots was 132 (City of Bombay) Squadron. James Hyde and Collins Joseph from Trinidad flew alongside Arthur Weeks of the Barbadian Contingent. Beautifully composed press photos of Weeks and Hyde with the squadron mascot, a shaggy dog named Dingo, suggest that the RAF was keen to promote the idea of the Empire coming together in Britain's defence. 132 Squadron was formed in Peterhead in north-east Scotland, but soon relocated to Newchurch on Romney Marsh in Kent, to join the hugely dangerous 'Rhubarb' missions into Occupied Europe.

Weeks survived the war but Joseph was hit by friendly fire on the last day of 1944 over Malmedy on the Belgian-German border. Hyde's Spitfire IX was shot down as he provided aerial cover for the British paratroopers fighting to capture the Rhine river crossing at Arnhem in the Netherlands.

The attrition rate for Spitfire pilots was high. New recruits, particularly those with previous flying experience, were desperately needed. Fortunately, toward the end of the Battle of Britain, more and more American pilots were making their way across the Atlantic. A decision was made to group them in new 'Eagle Squadrons'. All-American squadrons fighting for Britain would pack a powerful propaganda punch. Sure enough, magazines, newsreels and movie-makers queued up to grab a slice of the Yanks flying Spitfires. Like the GIs who would eventually follow them, the squadrons brought American candy and nylon stockings to raise spirits and light a beacon of hope that the US military might ride to Europe's rescue once more.

What the new squadrons didn't initially see was much action. By the time the first of the three Eagle Squadrons was fully operational, the Battle of Britain was over. Art Donahue was transferred to 71 Eagle Squadron with the other early arrivals from America – Eugene Tobin, Andrew Mamedoff and Vernon 'Shorty' Keough – but he was so frustrated by the peace and quiet that he requested a switch back to a non-American squadron.

It was a move that certainly gave him all the action he was looking for, taking him to Singapore for the desperate fight against Japan's lightning invasion. It proved to be one of the RAF's most disastrous campaigns of the war. The island's air-raid warning system was fatally flawed and dozens of Brewster Buffalo and Hawker Hurricane fighters were destroyed on the ground by Japanese bombers. By the time the Imperial ground forces were within sight of Singapore, Japan had complete air supremacy. Art escaped on one of the last transport planes out of Singapore.

By summer 1941, the Eagles had seen their Hurricanes replaced with the latest Spitfire Vbs and began to engage in significant offensive operations against the Germans. This culminated in their starring role over Dieppe in August 1942, when all three Eagle Squadrons contested one of the fiercest aerial battles of the war.

Supporting a poorly planned attack by Canadian infantry and British commandos on the port of Dieppe, the Spitfires of 71, 121 and 133 Squadrons flew sortie after sortie against relentless Luftwaffe attacks. The invading force intended to briefly occupy the town and destroy the port facilities before mounting an orderly retreat, but fierce German fire had pinned the troops and tanks to the landing beaches. In the ensuing chaos the Eagle squadrons claimed to have downed two Ju 88s, five Fw-190s and a Dornier Do217. Six Eagle Spitfires were lost alongside another 100 Allied aircraft. Around half of the invading infantry force was killed or captured by German troops.

One month later, the Eagle Squadrons were abolished and their pilots transferred to American squadrons. Japanese aircraft had attacked the US naval base at Pearl Harbour in Hawaii on 7 December 1941 and America had entered the war against the Axis powers. The United States Army Air Force was already flying combat missions from England when, on 29 September 1942, Air Chief Marshall Sholto Douglas, head of RAF Fighter

Command, addressed the pilots and crew of all three Eagle squadrons:

> The US Army Air Corps – their gain is very much the Royal Air Force's loss. The loss to the Luftwaffe will no doubt continue as before. In the eighteen months which have elapsed since your first unit became operational, Eagle pilots have definitely destroyed some 73 enemy aircraft. It is with great personal regret that I today say goodbye to all you boys whom it has been my privilege to command. You joined us readily and of your own free will at a time when our need was greatest and before your country was actually at war with the common enemy. You were the vanguard of that great host of your compatriots who are now helping us make these islands a base from which to launch that great offensive which we all desire.[18]

This new alliance of the RAF and the United States Army Air Force, fighting together against the Nazis, was neatly illustrated by a photograph that appeared in newspapers across the United States. 'Pretty brunette' WAAF officer Dinah Baxter was pictured outside Holy Trinity Church, Southall, with her new husband, beaming beneath his brand new USAAF cap. Stocky Texan, Forrest Dowling, had trained on Hurricanes in Canada before sailing for England. Here he spent six months flying combat missions with 71 Squadron until his Spitfire, 'Miss North Dallas', was shot down in a fierce battle with four Messerschmitt 109s. Just over a year later another photograph appeared in the British press. On his knee, Forrest was bouncing a chubby blond baby boy as Dinah tickled their son's feet for the camera. He was the first baby of the war born to an American pilot and a British officer. The newspapers christened little Michael as 'Britain's first Eaglet'.[19]

Most American pilots willingly transferred from the RAF to the new USAAF units, but some experienced a momentary culture shock, as Florida racing driver Colonel Reade Tilley told the BBC in 1976:

> It was considerably different. In the RAF mess one dined like a gentleman; white table cloths, silver tankards, mess stewards with white jackets. You didn't find this in the American air force but you did have peanut butter on the table.[20]

Art Donahue wasn't there to taste the peanut butter. He had died just three weeks earlier. At dawn he had flown his Spitfire Vb alone to the Belgian coast, deliberately flying at the speed of a bomber to attract the attentions of a German nightfighter. The ruse worked but the heavily armed Junkers Ju 88 held its own and the Spitfire crashed into the English Channel. Art's body was never recovered.

Eugene Tobin and Vernon Keough had already been killed in action. Tobin was shot down over Boulogne in a dogfight with Messerschmitt 109s, while Keough was last seen chasing a Heinkel 111 bomber away from a convoy off Flamborough Head in Yorkshire. Andrew Mamedoff followed his friends one week after the handover – his Hurricane smashed in foul weather against an Isle of Man hillside. These young Americans, in search of adventure, all met their deaths in defence of a country that was not their own.

When he first arrived in England, Art had written to his parents back home with a prediction: 'My life may not be long, but it will be wide.'[21]

Art's account of his first weeks in England added further lustre to the international reputation of the Spitfire. For Americans, whatever their view of the war, it had become the symbol of British resistance. The Germans feared it above all their airborne enemies – captured air crew would often claim to have been a shot down by a Spitfire, suggesting that they'd been defeated by a worthy opponent rather than the Hurricane or anti-aircraft fire that may well have brought them down. It's possible to pinpoint a moment in the war when this transition began – the day when the Spitfire stopped being just another fighter and became a legend. It was a day that turned a battle and, ultimately, a war.

20

Plotting the Future

As the sun rises on 15 September 1940, operation rooms across southern England buzz into life. These are the nerve centres of Fighter Command. They receive reports of enemy aircraft approaching and guide Spitfire pilots to their targets. The past three weeks have been tough, with relentless attacks on airfields and the first mass bombing raids on London.

The operation rooms are staffed by WAAFs – elite members of the Women's Auxiliary Air Force – smart young women like Peggy Balfour. Peggy's exactly where she wants to be. On the day that war was declared, she ran to the primary school where she taught and asked the Headmistress if she could manage without her. She was, she said, going to 'join the Air Force right away'. Three weeks later she had enrolled at Farnborough.

> During my interview I was asked what I should like to do in the WAAF. Just anything to do with aircraft, I told the senior WAAF officer. We chatted for quite a while, then she told me she thought I should go into Fighter Command as a plotter and work in a Sector Operations Room. I had no idea what that meant. Plotting? Anything to do with spying? Surely not! But

Fighter Command sounded fine. I signed up for that at once.[1]

Peggy plays a critical role in the defence of Britain; she's part of the Dowding System, the world's most technologically advanced air raid early warning system.

New radar stations stretching from the Orkney Islands to the Isle of Wight, known as Chain Home, detect formations of German bombers as soon as they take off from their bases in France. Groups of 100-metre-tall transmitting towers, with fourteen sets concentrated in south-east England between the Wash and the Solent, send out broad beams of radio pulses. If the pulse encounters aircraft, a signal is returned to the shorter receiving towers. That information is interpreted by highly skilled operators hunched in front of cathode-ray screens. Their estimates of the size, altitude and directional heading of enemy formations are called through on secure private phone lines to Fighter Command's filter room at Bentley Priory in northwest London. A plan of the emerging battle is prepared and then relayed to Group and Sector operation rooms, where plotters like Peggy mark the movements of the aircraft on a huge table-top map of south-east England. Sector Station Commanders look down on these maps and decide when and where to deploy their precious Spitfire and Hurricane fighters. The Sector Stations receive updated information as it becomes available and use radio to direct fighters already in the air.

It's not a foolproof system. Low-flying aircraft go undetected and the radar stations can only look out to sea. Once the bombers cross the coastline their movements must be tracked by more primitive means – the volunteers of the Observer Corps armed with binoculars. Interpreting the radar signal is an art, not an exact science. Pilots are often given inaccurate information about the height and composition of enemy formations. Despite those drawbacks, the Dowding System – named after its energetic promoter and head of Fighter Command, Hugh Dowding – does give the RAF an enormous defensive advantage. It's a fully integrated defence system that produces a single comprehensive view of the airspace, edits that information and cascades it down to the crews manning anti-aircraft guns, searchlights and barrage balloons as well as the fighter aircraft on the ground and in the air. Pilots can be sent where and when they're needed, spending their precious air time in combat rather than on speculative patrols. It sounds straightforward but this innovative approach to national defence requires effective data handling and crystal clear verbal communication from hundreds of operators.

Peggy's particular job demanded intense concentration. At the height of a raid there would be a stream of data coming through her Bakelite earphones. The information had to be interpreted and then plotted, with a croupier stick used to push small wooden discs representing squadrons around the map. Prospective recruits to Operation Rooms were asked if they could knit and talk at the same time.

Peggy was well aware of the importance of her work. She'd just met a boy who flies a Spitfire.

> A week or two after we arrived I went to a Camp dance held in the Sergeant's Mess, it seemed the thing to do apparently, so I went along.[2]

RAF Digby in Lincolnshire – 'rather far north' – according to Peggy's diary, is her first official posting.

After a fairly dreary few weeks of initial training at West Drayton – 'not the loveliest of places' – all marches, inoculations and parades, Peggy had been sent to a stately home near Leighton Buzzard in Bedfordshire to learn to plot. This was much more like it, plotting imaginary raids and learning the lingo of the RAF. Issued with a smart new uniform, Peggy took the train to RAF Digby, arriving on 8 December 1939 in plenty of time for a stimulating round of pre-Christmas fun:

> I had never been to a dance before without an escort – and without really knowing who would be there or what it would be like. I stood in the doorway of the hall – in the shadows, taking stock of my surroundings before venturing in. It all looked great fun. The music was playing merrily and everyone seemed to be danc-ing happily. I remained where I was. Then someone came over and asked me to dance with him. What joy – he was tall, and I found danced beautifully. I loved dancing.[3]

Walter Lawson was known as Jack to his family and 'Farmer Lawson' to his squadron mates, because he often spoke of being a farmer after the war. He joined the RAF in 1929 at the bottom of the rank structure, starting as an Aircraft Apprentice, becoming a Fitter in 1931, a trainee pilot in 1936 and finally a commissioned officer just before the Battle of Britain. Jack liked to dance, and that night, it seems, he found the perfect partner:

> We danced and chatted and then I remember, he called across to one of his friends, 'Tubby, guess what, I'm dancing with a WAAF who's asking about

centrifugal force.' I HAD asked him. I wanted to know. I had been reading an article – which I didn't understand – about it in my 'Times' that morning in Ops. I got to know my first dancing partner quite well, Jack Lawson. I found he never fussed me and danced quite beautifully – completely to my liking. And above all he was great fun and made me laugh.[4]

It had been a long, tortuous, barely trodden path to the Spitfire cockpit for Jack Lawson. The RAF had traditionally recruited its pilots from the upper class. Before the war it was unusual to reach pilot officer status without attending one of the top public schools, but by 1936 it was clear that Germany's military build-up required a rapid response. The net of recruitment had to be widened and a new breed of 'sergeant pilots' was taken on through the Royal Air Force Volunteer Reserve. These were generally young men with technical experience but none of the connections required to enter the service as an officer cadet.

Despite the influx of new blood, the class divide was still rigidly maintained at the outbreak of war. There were separate messes for officers and each was assigned a batman, a personal servant, to make their morning tea and keep their more comfortable accommodation and smarter uniforms looking tip-top. At Jack's 19 Squadron there was even an unspoken agreement that officers and sergeant pilots drank in different pubs. Jack's blunt Yorkshire comrade, Flight Sergeant George Unwin, remembered that: 'You never mixed at all, you were trained like that, the RAF built it into you.'[5]

Jack, then, was an unusual man, his drive and talent taking him from the lowest of RAF ranks to the cockpit of a Spitfire. Peggy knew that she had found somebody special:

> When Jack Lawson was free he used to send notes to the Guard Room, asking when he could meet me. It was all fun. Jack had a car, so we were able to leave the Camp. We used to visit the local pubs, or buy fish and chips and eat them out of newspaper – it had to be newspaper I remember.

The weather always seemed to be fine and sunny that spring, as they drove the arrow-straight roads of Lincolnshire through fens, fields and market towns. Jack and Peggy talked, laughed and argued:

> I remember, he was always teasing me about my ideas and, I think, thought me a little mad. But we enjoyed our jaunts into the countryside. We had lots of fun and, best of all, Jack never fussed.[6]

As their relationship deepened, war flared up all around them. The Battle of Britain began and Jack's squadron was moved south from Digby to better counter the German threat. Day after day Jack flew his Spitfire, attacking bombers set on destroying Britain's defences – ports, radar stations, aircraft factories and the very airfields where Peggy and her fellow plotters tracked the battle:

> June and July were very busy months in Ops. I received letters from Jack, written in haste, telling me how the squadrons moved around – how once his aircraft had been shot up and how lucky he'd been. And how, at one airfield they visited they were able to make use

of the swimming pool there – lovely during the hot weather – but only once – the next day – nothing, it had been blasted to rubble, and they were off again.[7]

German raids were taking their toll. Spitfires were being destroyed on the ground and in the air. For the pilots, each day was a personal battle against stress and exhaustion:

In a letter to Jack, arguing and complaining about red tape and discipline, I must have sounded gloomy. He wrote back telling me off – said it made him gloomy and sad also. I never did that again. In no way could I become an additional burden. I thought flying must need 100%. Life was difficult enough just staying alive from day to day – that was enough.[8]

From dawn to dusk, Jack was on stand-by, ready to fight at just a moment's notice.

In one letter Jack had to break off to 'mount aloft', as he put it. Back again, he told me he had caught nothing – 'better luck next time'. I used to worry and his letters made me think. There seemed to be so little respite and life was so fragile. However, nothing was gained from being miserable. Whatever happened in those days, we always kept laughing, and making as much fun for everyone as we could. That way we all got by. And the summer passed.[9]

On 5 September 1940, Jack led Black Section on a 15,000 feet patrol over Hornchurch. He spotted forty to fifty Dornier bombers protected by fifty fighters. He threw his Spitfire into the bomber formation and let off a two-second burst of his eight Browning machine guns. A Dornier dived away, apparently out of control, but the rear gunner sent an arc of tracer fire back toward Lawson. He was hit and broke off the chase. As Lawson headed unsteadily for home, he found himself:

> on the tail of an Me109 at which I fire a short burst at about 300 yards causing him to go into a vertical dive. I was again fired at from astern so did another steep diving turn to the right and could not see what happened to the Me109. My tailplane and port wing were damaged so I returned and landed at base.[10]

Jack's time in the air had lasted just over an hour, but death had been dealt out and narrowly avoided twice. There was good news waiting for him when he landed – a promotion to Flight Commander, in charge of one of 19 Squadron's two flights and, much more appealing, two precious days to celebrate with Peggy:

> It was the only leave Jack and I had at the same time – ever. The weather was lovely, the sun shone and it was warm and mild. At this time we had no car, we went for walks and visited local pubs – the countryside outside the camp was lovely just then. We talked and argued and laughed.[11]

By September, Adolf Hitler was losing patience. The invasion of Britain was pencilled in for 20 September and must be launched before unpredictable winter weather made the crossing of the English Channel too dangerous. Recent raids on and around London had met limited resistance and senior figures in the Luftwaffe believed that their policy of targeting airfields and aircraft factories had achieved its goals. The RAF was on its last legs. Hitler told his commanders that five more days of good flying weather would be enough to wipe out Fighter Command. Just a few more raids would clear the Spitfires and Hurricanes from the skies.

*

The RAF's troubles hadn't all been inflicted by the Luftwaffe. More Spitfires were desperately needed and dead and injured pilots had to be replaced urgently, but there was also a tactical crisis in play – a catastrophic clash of personalities at the top of Britain's air force. It pitched a cautious tactician against a reckless national hero.

British tactics against the Luftwaffe had, until now, been simple. When a formation of German bombers was spotted gathering over France or crossing the English Channel, two squadrons of around twenty fighters would typically be scrambled to intercept them before they reached their targets. This damage limitation strategy was now being challenged by a theory known as the 'Big Wing'.

Five squadrons – around sixty fighters – would take to the air. It didn't matter if they met the Germans before or after they'd dropped their bombs. The priority was to shoot down as many enemy planes as possible.

This Big Wing theory was promoted by the commander of 12 Group, the sector of Fighter Command protecting the Midlands. Trafford Leigh-Mallory had watched with mounting frustration the cautious tactics of his counterpart protecting London – the head of 11 Group, Keith Park. While Park's tactics may be limiting the destructiveness of German raids, too many RAF fighters were being lost and too many Luftwaffe pilots and crew were making it home safe and well. Crucially, Leigh-Mallory had a charismatic supporter, an influential Squadron Leader at RAF Duxford, a powerful personality who inspired love – and hate – in equal measure: Douglas Bader.

Douglas Bader joined the Royal Air Force as a cadet, directly from his Oxford public school. He was commissioned as an officer in 1930, his precocious flying skills bringing him to the attention of the RAF's aerobatic team. He quickly became something of a star with a show-stealing performance at the 1931 Hendon air show, catching the eye of the newspapers. Self-confidence slid effortlessly into arrogance and a few months later he attempted a particularly dangerous stunt – a slow roll at low altitude in a Bristol Bulldog fighter. It was a manoeuvre expressly banned below 1,000 (300m). Bader reportedly attempted it at a flying altitude of just thirty feet (9m). His wing hit the ground and Bader crashed spectacularly. In his logbook he wrote: 'Crashed slow-rolling near ground. Bad show.'

The twenty-one year old pilot had both of his legs amputated, but, with a new set of prosthetic limbs, he was determined to get back into the cockpit. The RAF initially refused to return him to a front-line role, but, as he told the BBC in 1978:

> You've got to be positive about it you see. My attitude is I'm not disabled and I'm going to kick these bloody legs around the place and make them do what I want.[12]

With the outbreak of war, official resistance to Bader crumbled and he swiftly proved his worth as an aggressive fighter pilot and an inspirational leader. Firmly back in the cockpit, he was given command of a squadron at RAF Duxford near Cambridge. This was his chance. Bader was itching to try out the Big Wing theory in the Battle of Britain, much to the irritation of the man in charge of the RAF fighters defending London – Keith Park:

> We were fighting at 11 Group with our backs to the wall. This wasn't a little game. We were not in a position to try out tactical theories of this and that, big wings or little wings. We were fighting for our existence, for the existence of London and the Empire.

Interviewed twenty years later, Park remained absolutely secure in his conviction – this was no simple clash of personalities, no minor disagreement over strategy:

> Had I tried to adopt Bader's theories of the Big Wings of five squadrons, I would have lost the Battle of Britain and this would have resulted, we know now, in an invasion by sea and by air by Hitler with his armoured divisions at a time when our Army hadn't enough rifles to go around, leave alone tanks and anti-aircraft guns. I might well have lost the war for the Allies.[13]

By September, Park's patience was running thin. His 11 Group had so far borne the brunt of the battle. Bader and the 12 Group squadrons were frequently asked to protect 11 Group aerodromes during large-scale raids, but the Big Wing took so long to form up at the correct altitude that Park's airfields were often hit, and the Germans bound for home, long before 12 Group aircraft were in a position to engage them. As far as Park was concerned, this rogue system was putting his whole strategy, and the successful operation of Fighter Command, in jeopardy.

*

There's a tremendous clatter and roar of Daimler-Benz and Jumo engines as hundreds of German fighters and bombers are readied for action on the coast of France. It's dawn on 15 September 1940.

In England two rival camps prepare to face them. To the south, 11 Group, controlled from RAF Uxbridge, is ready to use the familiar tactics of fast and targeted response. To the north, 12 Group, including RAF Duxford, is committed to the Big Wing approach – attacking German intruders with overwhelming force. Those tactics are about to be tested, along with the courage of WAAF plotter Peggy Balfour and Spitfire pilot Jack Lawson.

Prime Minister Winston Churchill senses that Sunday the 15th will be a big day. There had been significant raids over London on Saturday and the forecast for today is ominously clear and sunny. With his wife, Clementine, he's driven 25 miles south from his official country home at Chequers to Uxbridge on the western edge of London. He arrives at 10.30, to be met by Keith Park. Park reminds the Prime Minister of the air conditioning fifty feet below. He mustn't light his cigar in the bomb-proof Operations Room.

When Churchill arrives, the room is peaceful and the map empty. He takes his place in the 'Dress Circle' overlooking the Ops table. The latest weather report shows an area of high pressure off the Bay of Biscay and an associated trough of low pressure off the south coast of Norway. The wind over England is coming from the north-west and it's steadily rising. At 20,000 feet that should put a powerful nose wind in the face of any bombers taking the short route from Calais to London. That will provide Fighter Command with a few crucial extra minutes to get squadrons airborne and in position. Park warns the expectant Prime Minister that things may turn out to be quiet today. Park doesn't really believe that, and after just a quarter of an hour the action begins. The last four major daylight raids have targeted London. Park sees no reason to expect a change of strategy, so defences have been arranged accordingly

Radar has already picked up '30 plus' gathering over Dieppe. Opposite Churchill is a giant blackboard marked with six columns for each of 11 Group's Sector Stations. Bulbs light up on this display panel as squadrons are brought to stand-by. Reports come in of '20 plus' and '40 plus' , even '80 plus' assembling over the coast of France. It's obvious that a major raid is on its way. The first British fighters – 72 and 92 Squadron Spitfires from Biggin Hill – are scrambled and sent to 25,000 feet – nearly 8km – high above the German fighter cover.

Churchill was a war correspondent long before he was a politician. He relishes his moments as a first-hand witness to great events. He writes notes as he watches from his glass booth, looking out over the map table, the blackboard and the twenty or so young women plotting the progress of wave after wave of German bombers:

> Presently the red bulbs showed that the majority of
> our squadrons were engaged. A subdued hum arose

from the floor, where the busy plotters pushed their discs to and from in accordance with the swiftly changing situation. Air Vice Marshall Park gave general directions for the disposition of his fighter force, which were translated into detailed orders to each Fighter Station by a youngish officer in the Dress Circle, at whose side I sat. He now gave the orders for the individual squadrons to ascend and patrol as the result of the final information which appeared on the map-table.

Park paces the floor as the Luftwaffe squadrons are identified, positioned and challenged:

The Air Marshall himself walked up and down behind, watching with vigilant eye every move in the game, supervising his junior executive band and only occasionally intervening with some decisive order, usually to reinforce a threatened area. In a little while all our squadrons were fighting, and some had already begun to return for fuel. All were in the air. The lower line of bulbs was out. There was not one squadron left in reserve.[14]

The first raiders reach the English coast near Dungeness at around 11 in the morning. Dornier 17 bombers with a large escort of Messerschmitt 109s appear to be heading for the capital. Spitfires engage the escort over Kent, but the bombers roar onward, following the pathway of the Thames toward London. 12 Group's Big Wing is ordered into the air. Jack Lawson and his comrades head south. It will be the first time that the Wing goes into action with its full strength of five squadrons.

Despite around 250 British fighters in the sky, the German formation reaches south London relatively unscathed. They target Latchmere Junction, an important railway link, connecting lines from north-west London to Clapham Junction and the terminus at Waterloo. The rail viaducts and many homes in Battersea and Wandsworth are hit. A small group of Dorniers press on northwards.

This will be a day long-remembered by pilots from both sides. Sergeant Ray Holmes flies his Hurricane toward the formation as they close in on central London:

> The Dorniers didn't fly particularly tight which was to their disadvantage. If they had done they'd have had better firepower to beat off the fighters. But our CO went at them in at a quarter attack and more or less went through them and spread them out a bit.[15]

Spotting three bombers split from their pack, Sergeant Holmes turns toward them.

> I made my attack on this bomber and he spurted out a lot of oil, just a great stream over my aeroplane, blotting out my windscreen. Then as the windscreen cleared, I suddenly found myself going straight into his tail. So I stuck my stick forward and went under him, practically grazing my head on his belly.

Holmes is lucky – the bomber had aimed a flame-thrower at his plane. At low level this could have been a deadly weapon, at high altitude the flamethrower failed to ignite so he was hit by a stream of oil not fire. He follows the bombers over the city, their desperate struggle watched by fascinated spectators on the ground. He attacks the second Dornier:

I got to the stern of the aeroplane and was shooting at him when suddenly something white came out of the aircraft. I thought that a part of his wing had come away but in actual fact it turned out to be a man with a parachute coming out.[16]

Travelling at 250 mph, Holmes has no time to take evasive action. The parachute gets caught on his starboard wing, with the petrified German hanging straight out behind him. The plane is dangerously unbalanced:

> All I could do was to swing the aeroplane left and then right to try to get rid of this man. Fortunately, his parachute slid off my wing and down he went, and I thought, Thank heavens for that!

As Ray Holmes struggles to shake off the German airman, the third bomber continues, apparently unperturbed, in the direction of Buckingham Palace. Holmes accelerates, gets in front of the Dornier, and prepares a head-on attack:

> As I fired, my ammunition gave out. I thought, 'Hell, he's got away now'. And there he was coming along and his tail looked very fragile and very inviting. So I thought I'd just take off the tip of his tail. So I went straight at it and hit his port fin with my port wing.

It's an audacious manoeuvre, trying to knock the tail off a sturdy bomber with a flimsy little fighter:

> I didn't allow for the fact that the tail fin was actually part of the main fuselage. Although I didn't know it at the time, I found out later that I had knocked off the whole back half of the aircraft including the twin tail![17]

The third Dornier is torn apart by the collision, but Holmes's plane is also fatally damaged. He bails out and floats to the ground near Victoria Station, where he's greeted by cheering crowds. The battle is caught on film and Holmes will be feted as the hero who saved Buckingham Palace.

The pilot of the Dornier, Robert Zehbe, had already bailed out – in fact there may have been no surviving crew left on the plane as Holmes rammed it. Zehbe lands close to the Oval Cricket Ground, where he's set upon by an angry mob. A Home Guard unit pulls him from the bloodthirsty crowd, but he dies of his injuries the next day.

Holmes's fighter crashes into the junction of Buckingham Palace Road and Pimlico Road with such force that most of his aircraft is buried. It was excavated in 2004 and the brass firing button, set to FIRE, was presented to a delighted Ray Holmes.

Back in the air, the Duxford squadrons arrive. Jack Lawson, Peggy's boyfriend, is part of Douglas Bader's Big Wing. The five squadrons arrive late to the battle, but Jack makes up for lost time. His combat report captures the relentless action of the day:

> I was leading 19 Squadron in The Wing formation. We sighted 20 Dornier 17s escorted by Me109s. We were flying at about 25,000 feet, they were at about 15,000. I went on ahead of the bombers and turned to deliver a head on attack. I fired a burst at about 400 yards, diving past the left hand Do17 . . . I then turned around and attacked the same aircraft from the rear, opening fire at 300 yards and closing slowly to about 50 yards. I saw pieces falling off from the starboard

wing and his starboard engine started streaming glycol. He drifted away from the main formation and started to glide down with glycol still streaming from his starboard motor. I then started after the remainder of the formation. I got a very short burst on another Do17. My ammunition ran out so I returned home.[18]

The first German wave retreats. Despite relentless attacks from 11 and 12 Group, most of the tough Dorniers, led by a hugely experienced formation commander, Major Alois Lindmayr, have survived to fight another day. They turn for France, sped home by a strong tail-wind.

But this was just the start. By two o'clock, the second, bigger wave – estimated at around 150 bombers and 350 fighters – is on its way to attack London's docks. More than 250 British fighters are scrambled to meet them.

Hans Zonderlind, front gunner of a German Dornier bomber, like many Luftwaffe flight crew that day, is horrified to see so many British fighters ranged against them:

We saw the Hurricanes coming toward us and it seemed that the whole of the RAF was there, we had never seen so many British fighters coming at us at once . . . All around us were dogfights as the fighters went after each other, then as we were getting ready for our approach to the target, we saw what must have been a hundred RAF fighters coming at us . . . where were they coming from? We had been told that the RAF fighters were very close to extinction.[19]

Despite the imposing sight of so many fighters, the escorting Messerschmitt 109s effectively screen the bombers and most of the formation reaches London intact, ready to line up its targets – Surrey Commercial Docks, West India Docks and Royal Victoria Docks. By now, though, most of London is enveloped in low cloud. It's increasingly difficult to spot any definite targets, so bombs drop randomly across the city.

The return of the Big Wing, now refuelled and rearmed, adds to the impression of Fighter Command bristling with resources and resolve. In fact, the sheer number of aircraft in the sky creates a general sense of confusion, with up to sixty aircraft at a time getting in each other's way as they converge on the retreating Luftwaffe. As the Duxford Big Wing attempts to intercept a stream of bombers, they're bounced by escorting Messerschmitt 109s. Sergeant George Unwin is flying with Jack Lawson as part of 19 Squadron:

> I got lost from my squadron and when I looked, there was not a soul in the sky. Then in the distance, I saw some ack-ack and I went toward it and I saw these waves upon waves of German bombers coming in. There seemed to be hundreds of them and I forgot about anything else. Suddenly I found aeroplanes whizzing all around me. I'd flown smack into the middle of the fighter escort. Damn fool![20]

Jack Lawson escapes the mêlée and chases a fighter over the Channel, but the Me 109 pivots and switches from prey to hunter, forcing him out of the battle. In his logbook that night, Jack describes how he nursed his battered Spitfire back to RAF Fowlmere, ending the entry with a simple, heartfelt, 'Thank God'.

Fifty-six German aircraft were shot down that day, while the RAF lost twenty-six. London's docks were hit hard and the Luftwaffe still had plenty of planes in reserve, more than enough to continue the attack into the night, battering the East End of London.

If 15 September was a victory for the RAF, it was a minor one, but it was a minor victory that soon proved to have enormous consequences.

The Royal Air Force was quite clearly still in daytime control of the skies over England. As far as Hitler was concerned, that made an invasion impossible. On 17 September, Operation Sea Lion was postponed indefinitely. The next day 200 British bombers attacked the German barges and army camps on the French coast. The invasion fleet was dispersed and German attention turned east to the Soviet Union.

Jack Lawson and Peggy Balfour both survived the day, but a year later Jack's Spitfire was shot down by a Messerschmitt 109 over Rotterdam Harbour as the Royal Air Force took the battle into Occupied Europe.

Peggy lived until 2000, sixty years after the Battle of Britain. She never married and she never forgot Jack Lawson or one particular heart-stopping moment when he revealed his true vulnerability.

Jack had told her of his pleasure in the simple things of life, the joy of flying and of his one and only fear – that terrible moment when his Spitfire would be hit by tracer fire, when fuel would gush into the cockpit, when he would frantically push and pull the hood of his cockpit, trapped as the fuel burst into flame around him:

I couldn't move or speak. Then Jack turned and just said, 'Remember, whatever happens, it is always in God's hands.'

The spell was broken, the picture faded. 'Yes,' I said, 'I know that and I will remember always. All the same, take care.'

And standing there, suddenly I knew. No one else will ever take Jack's place. For me, that's it for ever. I will not let him know – not yet, we are all too busy. I don't want him to think about anything but flying. Everything else must wait.[21]

15 September 1940 was not the end of the Battle of Britain. It certainly wasn't the end of the war. Bigger, darker stories waited to be told at Stalingrad and Dresden, Auschwitz and Dachau. What 15 September showed was that defeat was not inevitable. The Luftwaffe, which had swept all before it since the invasion of Poland a year earlier, could be turned back. Tyranny need not prevail.

The symbolic value of the day was quickly recognised. As early as 1943, the date was commemorated as Battle of Britain Day and the Ministry of Information rushed to produce a fold-out illustrated brochure that promised to explain to the British public 'How Britain Won the Greatest Air Battle of History'. An official government history of the Battle of Britain followed. Priced at 6d, the first edition of two million copies sold out. Puffin Books produced a colourful account of the conflict for children, describing how: 'That battle overhead, fought by so few, saved us from being enslaved like the French and the Poles.'[22]

On the Puffin book jacket, what else but a Spitfire soaring upwards?

The cost of those symbolic Spitfires rose exponentially after 15 September. A stroll through the churchyards of Southampton reveals the price paid. There are two dates carved on so many gravestones – 24 and 26 September 1940 – the two deadly bomb attacks on the Supermarine factory.

In Hollybrook Cemetery there's sixteen-year-old Apprentice, Mervyn Hawkins; in the West End Old Cemetery lies eighteen-year-old Aircraft Electrician Ronald Barfoot. And in St Mary Extra, so many names; twenty-year-old Aeronautical Draughtsman Leonard Cox, 'Shop Boy' Douglas Cruikshank, who was just fourteen years old, and, of course, Margaret 'Peggy' Moon – Clerical Typist and a good friend to Joan Tagg and Cyril Russell.

For the surviving workers of Supermarine to recover from the tragedy of those lost young lives took great reserves of fortitude. It was a mighty effort of will that saw them dust themselves down and push forward one of the war's most audacious escape plans. An enterprise like that needs inspiration and powerful motivation, and that's precisely what was provided by the tenacious dogfighters of 15 September. Victory was possible and the Spitfire – and its young pilots, designers and builders – would lead the way.

Afterword

David Key checks the flimsy wedge of wood holding open the enormous steel door of the walk-in safe. It's quite an important bit of wood, apparently. If the door swings shut then it will take two or three men with the correct combination to get us out. That's assuming anybody knows we're here.

Today, Hursley House is a research and development hub for the computer company IBM. They've been here more or less since Supermarine left in the 1950s. Over 1,500 people work here, creating and developing the software that runs in the background of much of the technology we take for granted, from cash machines and share trading to the household objects we can control from our phones. Dave works as a software engineer here, but Hursley is also the home of his particular historical obsession – the designers and builders of the Spitfire.

Dave rummages in some cardboard boxes. They're stacked on a folding table he swears to be the original butler's polishing table. Apparently Lady Cooper's family silver was kept in this particular safe. He finds some copies of the *Ragazine* – the in-house magazine of the Supermarine Design team at Hursley Park. He's got copies from 1942, '44 and '46. These aren't just vintage pieces of corporate guff; they're written – and drawn – by the draughtsmen and -women themselves.

Not surprisingly, these people were pretty good artists and they clearly had no fear of offending the management. Each cover carries a cheeky cartoon of a senior staffer – there's one of Chief Designer Joe Smith puffing on his pipe as he ponders an absurdly over-engineered seaplane. Inside, the writers and artists needle the Germans, portray staff mindlessly flirting in the parkland, and skewer a particular manager renowned for his love of sleep and general laziness.

Sometimes it's the smallest thing that can give you a clue about the reasons for an organisation's success. Dave puts down the *Ragazine* and pulls out a packet of photos – all black and white. There's one of nine young women, smiling at the camera and holding rather elegant glasses of what might be port. Dave tells me they're girls from the Tracing Office. From the little statuette of a dancing couple on the table, it looks as though they're celebrating an engagement.

There's a team photo showing eleven women looking proud but painfully cold outside one of the Hursley hangars. They've got a leather ball and they're dressed in a random collection of shorts, skirts and the high-waisted, wide-cut trousers of the day. A football team maybe, perhaps netball? More photos to flick through – young bucks in cricket whites, intense women in lab coats, and the wedding portrait of tracer June Wild and her proud-as-punch draughtsman husband.

This is a community – they work, play and date here. They mock the management as much as the enemy. They're here at Hursley for one purpose – to maintain the Spitfire's position as the best fighter in the world – but they're young and bursting with life. They're going to enjoy themselves while they win the war.

Compare their lives and these photographs to the contemporary reports from Germany. Even before the war, their best engineers were nervous to speak out. If a design favoured by the regime proved impossible to build, few were eager to admit their failure. Minor setbacks resulted in the cancellation of important projects, while wild ideas were pursued at the whim of the Führer. Jews, Socialists and homosexuals had already been purged from the industry, and, as the war turned against the Nazis, every engineer and designer lived in constant fear of being called up to the Eastern Front.

When Supermarine replaced its trained Southampton riveters with enthusiastic shop girls from Trowbridge, Messerschmitt and Daimler-Benz were enlisting slave labour from Poland and Ukraine.

Dave remembers giving a tour of Hursley to Judy Monger, daughter of Ernest Mansbridge. He was one of R. J. Mitchell's right hand men in the Design Office in Southampton and a senior manager at Hursley. Judy asked to see his old office. This proved a little embarrassing, as it's now the Gents toilets. As a four-year-old, Judy had been with her father at Eastleigh airport in 1936 when the Spitfire prototype first flew. She vividly remembered visiting the factory and playing with the children of the other designers. Her husband, Gordon, joined Supermarine as a fifteen-year-old apprentice and worked on many different marks of Spitfire.

At the start of the war, Supermarine had a reputation as a tough employer. Trades unions were relatively weak and shop-floor workers could be dismissed with little or no notice. At Hursley Park, though, this really feels like a family business, forged in adversity and united in a common cause. Tasked with co-ordinating dozens of brand new factories and workshops and many more sub-contractors all across the UK, the Hursley team succeeded brilliantly. Enough of the right parts for the right planes arrived in the right places to keep the ravenous appetite of the Royal Air Force fed with Spitfires.

As the war ground on and the Luftwaffe's fighters fronted up with more muscle and more speed, the most creative of engineering minds had the space here at Hursley Park to think, the permission to fail and the resources to build again.

It's probably absurd to search for the soul of a machine, even one pulsing with memories and meaning like the Spifire. But as I look out over the soft green acres toward the English Channel, I can't help thinking about those ghosts that still fly with Squadron Leader Sugden in his Spitfire cockpit.

If they could stand with me now, those brave young people who built and flew the Spitfire, I like to think they would find a kind of contentment here, a sense of a job well done.

Appendix

The Spitfire Through Time

1895 *20 May* – Reginald Joseph Mitchell, designer of the
Spitfire, is born in Staffordshire.

1913 Noel Pemberton Billing founds an aircraft
production company at Woolston, on the River
Itchen, in Southampton.

1917 R. J. Mitchell travels south to work as personal
assistant to the new Managing Director of the
company now known as Supermarine. Small-scale
production of land planes designed for the wartime
military is superseded by the manufacture of a series
of flying boats, most designed by Mitchell.

1919 Mitchell appointed Chief Designer at Supermarine

10 September – Third Schneider Trophy contest;
Supermarine make their first entry in this
international competition for the world's fastest
seaplane. Mitchell watches and learns as the
Supermarine Sea Lion – and most other entrants –
fail to complete the Bournemouth course.

1922 *12 August* – Sixth Schneider Trophy contest; Henri
Biard pilots the Supermarine Sea Lion II to victory
in Naples, reaching a top speed of 150 mph.

1923 *28 September* – Seventh Schneider Trophy contest; Supermarine Sea Lion III is beaten into third place by two American aircraft. The Curtiss CR-3 sets a new world air speed record of 177.38 mph.

1925 *26 October* – Eighth Schneider Trophy contest; Mitchell's innovative Supermarine S4 single-winged seaplane is unveiled to the world at Baltimore, only to crash during trial runs for the competition. The United States win with a Curtiss R3C-2.

1927 *26 September* – Tenth Schneider Trophy contest; in a circuit of the Venice Lido, the Supermarine S5 wins, reaching 281.65 mph.

1928 Vickers take over Supermarine. This large engineering company has the money and expertise to back Mitchell's designs. Mitchell is contractually tied to the company and joins the Vickers board.

1929 *7 September* – Eleventh Schneider Trophy contest; Supermarine triumph again on home soil with Mitchell's S6.

12 September – Supermarine S6 breaks world air speed record, reaching 357.7 mph.

1931 *13 September* – Twelfth and final Schneider Trophy contest; Supermarine S6B wins the race on the Solent. As Britain has won three competitions in a row, she gets to keep the cup and the trophy is retired.

29 September – Flight Lieutenant Stainforth breaks the world air speed record, reaching a speed of 407.5 mph in the Supermarine S6B.

Mitchell begins work on a new fighter for the British government. The resulting aircraft, the Type 224, is an ungainly failure.

1933 *30 January* – Adolf Hitler is appointed Chancellor of Germany. Opposition is swiftly crushed and dictatorial powers seized.

14 October – Hitler withdraws from the League of Nations and the Geneva Disarmament Conference. Rearmament of the Luftwaffe accelerates.

1935 *3 January* – Air Ministry issues Specification F37/34; this authorises Supermarine to build a prototype fighter based on new plans proposed by R. J. Mitchell. Fitted with the Rolls-Royce PV12 engine, it will become the Spitfire.

1936 *5 March* – Maiden flight of the Spitfire prototype K5054. On landing, test pilot Mutt Summers tells the ground crew – 'I don't want anything touched'.

3 June – British Government orders the first 310 Spitfires from Supermarine.

1937 *11 June* – R. J. Mitchell dies of bowel cancer at the age of forty-two.

1938 *14 May* – Test pilot Jeffrey Quill flies the first production model of the Spitfire Mark I.

4 August – First Spitfire Mk I delivered to the RAF, 19 Squadron at Duxford.

1939 *1 September* – Germany invades Poland.

3 September – Prime Minister Neville Chamberlain declares war on Germany.

16 October – First combat for the Spitfire. Aircraft from 602 and 603 Squadrons shoot down Junkers Ju 88 bombers set to attack warships in the Firth of Forth.

1940   *10 July* – Battle of Britain begins. Small-scale nuisance raids by German bombers are followed by intensive attacks on shipping in the English Channel before the focus of the assault is turned on RAF airfields and air defence systems. Around 1,000 Spitfires are in active service with the RAF.

18 August – The Hardest Day; the peak of Luftwaffe efforts to knock out Britain's air defence systems. Both sides suffer heavy losses of aircraft. The Luftwaffe then turns its attention to the bombing of London and Britain's industrial cities.

15 September – Battle of Britain Day; heavy day bombing of London, repelled by RAF fighters. The success of the RAF persuades Hitler to postpone Operation Sea Lion, the planned invasion of Britain.

24 September – Heavy bombing raid on the Supermarine factories in Southampton. The bombs miss the factories but air-raid shelters are hit. Forty-two people are killed and sixty-three seriously injured.

26 September – Sixty Heinkel 111 bombers attack the Supermarine factories. Seven bombs hit the Woolston plant and one hits the neighbouring Itchen factory. Fifty-five people are killed. Production is halted and a plan to disperse Spitfire production is put into action.

31 October – Battle of Britain ends. Heavy daylight bombing has gradually declined for several weeks while night bombing of London and industrial cities has increased and will continue for many months.

23 November to 1 December – The peak of the Southampton Blitz. These heavy night-time raids cause significant damage to the mediaeval city centre, factories and port. The city would continue to be targeted sporadically until May 1944.

December – Maiden flight of the Spitfire Mark V. Intended as a stop-gap measure, over 6,000 of these are built, more than any other variant of the Spitfire. The main production version, the Mk Vb, is armed with two 20mm Hispano cannons, alongside four 0.303 Browning machine guns. It can reach 371 mph. Most are built at Castle Bromwich Aircraft Factory.

1941 *February* – The first of the Eagle Squadrons, staffed by American pilots in RAF fighters, becomes operational.

7 December – Japanese air force attacks the US fleet at Pearl Harbour in Hawaii.

11 December – United States declares war on Germany.

1942 *June* – First Seafires enter service; these Mk1Bs are Spitfire Vbs adapted for Fleet Air Arm use from aircraft carriers.

July – The Spitfire Mk IX enters service. Essentially a Mk V with a new Merlin 61 engine, this was a much improved aircraft with a top speed of 407 mph that could fly at a much higher altitude. More than 5,000 will be built.

19 August – The Raid on Dieppe; a disastrous attempt to seize the French port results in fierce aerial battles. Fifty-nine Spitfires are lost.

30 August – *The First of the Few* premieres at the Leicester Square Theatre, London. The highly romanticised story of the Spitfire's origins cements the golden reputation of Mitchell and his plane.

29 September – The Eagle Squadrons are disbanded, their pilots transferred to the United States Army Air Force.

1943 1 June – Leslie Howard, star and director of *The First of the Few*, dies on board a civilian airliner flying from Lisbon to Bristol, shot down by German fighters over the Bay of Biscay._

1945 8 May – Germany surrenders.

15 August – Japan surrenders, bringing World War Two to an end.

May 1948 to March 1949 – Arab-Israeli War; Spitfires fight Spitfires as the Egyptian and Israeli Air Forces attack each other. RAF Spitfires are destroyed in the air and on the ground in cases attributed to mistaken identity.

1949 28 January – The last Spitfire leaves the factory: a Seafire Mk 47 designed for use from aircraft carriers. It has a top speed of 451 mph.

1954 *1 April* – The last operational sortie flown by
an RAF Spitfire takes off from RAF Seletar
in Singapore as part of Operation Firedog in
the Malay Emergency conflict. It is a Mk XIX
photographic reconnaissance aircraft searching for
pro-independence Communist insurgents in the
jungles of the Johore area in the south of Malaysia.

Acknowledgements

This book would have been impossible to write without David Key's nose for a good story and his tenacity in chasing down leads from Supermarine workers and their families. As the site historian of Hursley Park and host of the superb Supermariners website, he's made absolutely sure that the designers and builders of the Spitfire aren't just remembered but celebrated.

Thanks also go to the many other people in and around Southampton who talked to us for the radio series and the book, most notably Alan Matlock, Geoffrey Wheeler, Steve Adams from the Bitterne Local History Society, Andy Jones and John Beck of the Solent Sky Museum and Rupert Rowbotham of the Nuffield Southampton Theatres.

The initial idea for the radio series, *Spitfire: The People's Plane,* came from the BBC World Service commissioning editor, Steve Titherington. My fellow producer, Emily Knight, and our researcher, Lisa Lipman, uncovered lots of the stories that made it through to this book and tracked down many of the most fascinating characters. Thanks are also due to the rest of the creative team who turned the initial idea into a beautiful piece of radio: our editor, Chris Ledgard, production co-ordinator Siobhan Maguire, the musicians of Public Service Broadcasting and our wonderful narrator, Tuppence Middleton.

Acknowledgements

The team at the Imperial War Museum, Duxford have been patient suppliers of detail throughout the process. Thanks to Emily Charles, Hannah Llewellyn Jones, Craig Murray and Adrian Kerrison. Many other museums and collections around the country supplied information and anecdotes but particular thanks go to Andrew Dennis at the RAF Museum in Hendon, to Sue Frost at the Market Lavington Museum, Richard Poad of the ATA Museum and to Sir Donald Spiers at the Farnborough Air Sciences Trust. The Battle of Britain Memorial Flight gave me the chance to get up close to their aircraft, pilots and engineers, so many thanks to Helen Fearn and Mark Sugden.

The work of specialist historians and writers has been invaluable, notably John Shelton on R. J. Mitchell, Calum Douglas on the rival engines of Germany and Britain, Giles Whittell and Sally McGlone on the ATA, Leo McKinstry on Castle Bromwich, Daniel Todman on Beaverbrook and Victoria Taylor on the Luftwaffe.

It's been an absolute pleasure talking to the friends and relatives of some of the most intriguing figures in the Spitfire story. Amongst those sharing memories, photographs and laughter have been Lou Lou Troup, Robin and Felicity Baker, Janine Thorp, Jo Denbury, Alan Ralph and our proud 'Eaglet', Michael Dowling. Thanks also to my family – Louise, Maddy and Archie – for supporting my long retreat into the 1940s with fresh ideas and vital insights into the teenage mind.

Finally, and most importantly, my grateful thanks go to all those extraordinary young men and women who built and flew the Spitfire and sacrificed so much in the desperate fight against Fascism.

Acknowledgements

Permissions Acknowledgements

Picture Credits

Author Images: p.1 top (courtesy of Southampton Cultural Services: Archives); p. 2 top (courtesy of the family of Dorothy Thorp, née Handel); p. 2 bottom (courtesy of Alan Ralph); p. 3 bottom right (courtesy of Robin Baker); p. 4 bottom (courtesy of Eileen Fowler); p. 5 middle (courtesy of Michael Dowling); p. 6 bottom (author's own photo)

Alamy: p. 6 top left (© Everett Collection, Inc.)

Getty Images: p. 1 top; (© Hulton-Deutsch Collection/Corbis/ Corbis via Getty Images); p. 3 top (© Capt. Horton/ Imperial War Museums); p. 3 bottom left (© Science & Society Picture Library); p. 5 top (© imagno); p. 7 top left (© A. Hudson / Stringer); p. 7 top right (© Tanner/Mirrorpix); p. 7 bottom left (© Fox photos / Stringer)

Imperial War Museum: p. 5 bottom; p. 6 top right; p. 8 bottom

Shutterstock: p. 4 top (© ANL/Shutterstock); p. 7 bottom right (© Roger-Viollet/Shutterstock); p. 8 top (© Bill Cross/Daily Mail/)

Southampton City Archives: p. 1 bottom

Text Permissions

Extract from *Spitfire: Portrait of a Legend* © 2008 Leo McKinstry reproduced by permission of John Murray Press, an imprint of Hodder and Stoughton Limited.

Extract, *More Memories of Bitterne and BLHS 44,* from the diary of Joan Tagg: © Joan Tagg, reproduced by permission of Bitterne Local History Society.

Extracts of the lyrics to *Spitfire Song* © The Christchurch History Society, reproduced by permission.

All audio extracts quoted from the Imperial War Museum are reproduced © IWM, by permission of their archives.

All copyright BBC radio archive material quoted is reproduced by arrangement with the BBC.

Interview extracts with Geoffrey Wellum from BBC1 documentary *Battle of Britain*: © 2010 Lion Television, reproduced by permission.

Acknowledgements

Interview extracts with Geoffrey Wellum from BBC1 drama-documentary *First Light*: © 2010 Lion Television, reproduced by permission.

Extracts from Geoffrey Wellum interview in the documentary *Spitfire* © Sea Blue Media Limited, reproduced by permission.

Extract of interview with Bette Blackwell © 2016 *Secret Spitfires* documentary by Howman & Cetintas.

Extract from Margaret Frost interview from BBC Radio Wales documentary *Miss Frost and the Spitfire* © 2015 Rondo Media Limited, reproduced by permission.

Extracts from audio interview with Stella Rutter © Solent Sky Museum, reproduced by permission.

Selected Bibliography

Beaver, Paul. *Spitfire People* (Euro Publishing: 2015).

Bishop, Patrick. *Battle of Britain: Day by Day Chronicle* (Quercus: 2013).

Brenton, Howard. *The Shadow Factory* (Nuffield Southampton Theatres: 2018).

Crompton, Teresa. *Adventuress: The Life and Loves of Lucy, Lady Houston* (History Press: 2020).

Donahue, Arthur. *Tally Ho! Yankee in a Spitfire* (Macmillan: 1942).

Douglas, Calum. *The Secret Horsepower Race* (Tempest: 2020).

Freudenberg, Matthew. *Negative Gravity: A Life of Beatrice Shilling* (Charlton Publications: 2003).

Howman and Cetintas. *Secret Spitfires* (History Press: 2020).

MacKenzie, S.P.. *The Battle of Britain on Screen* (Bloomsbury: 2007).

McKinstry, Leo. *Spitfire: Portrait of a Legend* (John Murray: 2007).

Mitchell, Gordon. *Schooldays to Spitfire* (History Press: 2009).

Nichol, John. *Spitfire* (Simon and Schuster: 2018).

Quill, Jeffrey. *Spitfire: A Test Pilot's Story* (John Murray: 1983).

Robson, Martin. *The Spitfire Pocket Manual* (Osprey: 2017).

Russell, Cyril. *Spitfire Odyssey* (Kingfisher: 1985).

Rutter, Stella. *Tomorrow is D. Day* (Amberley: 2015).

Sarkar, Dilip. *Spitfire! The Full Story of a Unique Battle of Britain Squadron* (Air World: 2018).

Saxon Childers, James. *War Eagles* (Tannenberg: 2016).

Shelton, John. *Schneider Trophy to Spitfire* (Haynes: 2008).

Warner, Carl. *Life and Death in the Battle of Britain* (Imperial War Museum: 2018).

Webb, Denis. *Never a Dull Moment* (J&KH Publishing: 2001).

Welch, David. *Persuading the People* (British Library: 2016).

Wellum, Geofffrey. *First Light* (Penguin: 2002).

Whittell, Giles. *Spitfire Women of World War Two* (Harper Perennial: 2008).

Notes

Introduction

1 *Ramblings*, BBC Radio Four, 14 November 2008.

1. The Perfect Target

1 'Not So Super Days', Joan Rolfe, *Bitterne Local History Society Local Paper No. 44*, 2018.
2 'J. Rolfe Remembers', *More Memories of Bitterne: Continuing the Patchwork of People and Places We Loved*, Irene Pilson (self-published, 1988), p. 287.
3 *Spitfire Odyssey: My Life at Supermarine, 1936–57*, C. R. Russell (Kingfisher Railway Productions, 1985), p. 11.
4 Ibid, p. 78.
5 *More Memories of Bitterne*, ibid.
6 Ibid.
7 Phil Pearce, a private memoir written 2018.
8 'Not So Super Days', Joan Rolfe, ibid.
9 *More Memories of Bitterne*, p. 289
10 Ibid, p. 289.
11 Ibid, p. 290.
12 Ibid.
13 *Spitfire Odyssey*, p. 74.
14 Phil Pearce, a private memoir written 2018.
15 'How I Survived the Bombing', Phil Pearce, *Bitterne Local History Society Local Paper No. 44*.
16 *More Memories of Bitterne*, p. 290
17 *Spitfire Odyssey*, p. 76
18 *More Memories of Bitterne*, ibid.
19 Ibid.

20 Imperial War Museum Sound Archive (IWM), 9849
21 *More Memories of Bitterne*, ibid.
22 Ibid, p. 291.
23 'Not So Super Days', Joan Rolfe, ibid.
24 IWM, 13855.
25 *Spitfire Odyssey*, p. 67.
26 *Spitfire*, BBC, first broadcast 6 March 1976.
27 Ibid.
28 BBC Home Service, 17 December 1940.

2. The Phoenix

 1 *Spitfire Odyssey*, p.89.
 2 Ibid.
 3 *Secret Spitfires: Britain's Hidden Civilian Army*, Howman and Cetintas with Gavin Clarke (History Press, 2020), p. 25.
 4 *Never a Dull Moment: At Supermarine, a Personal History*, Denis Le P Webb (J&KH Publishing, 2001), p. 140.
 5 Ibid, p. 141.
 6 Ibid, p. 140.
 7 Ibid.
 8 Ibid.
 9 Ibid.
10 *St James' Park: From Shirley Rec to Renovation 1907–2014*, Don Smith (Shirley Heritage Project, 2014), p. 39.
11 IWM, 7488.
12 'The Blitz', *Spartacus Education*, https://spartacus-educational. com/2WWblitz.htm
13 Ibid.
14 Ibid.
15 'Southampton in World War II', Jake Simpkin, http://www. jakesimpkin.org/ArticlesResearch/tabid/84/articleType/ ArticleView/articleId/10/Southampton-In-World-War-Two.aspx.
16 *The City of Coventry: A Twentieth Century Icon*, Adrian Smith, (I. B. Taurus, 2006), pp. 153–6.

3. From Hairdressers to Riveters

 1 *The Supermariners* blog, David Key, https://supermariners.wordpress. com/the-places/southampton/the-dispersal-1940-1941/

2 *Never a Dull Moment*, p. 154.

3 Ibid, p. 153.

4 *The Supermariners* blog, https://supermariners.wordpress.com/the-hutments-hursley-park/

5 Eileen Bell, as told to David Key, *The Supermariners* blog, https://supermariners.wordpress.com/the-hutments-hursley-park/

6 Ibid.

7 *Tomorrow Is D-Day: The Remarkable Story of Supermarine's First Draughtswoman*, Stella Rutter (Amberley Publishing, 2015), p. 121.

8 Ibid, p. 126.

9 *Secret Spitfires*, p. 91.

10 Ibid, p. 95.

11 'Spitfire: The People's Plane', episode 4, Alan Matlock, BBC World Service, 31 May 2020.

12 Essie Dean, as told to David Key, *The Supermariners* blog.

13 *Portsmouth Evening News*, 22 July 1943.

14 Janine Thorp, daughter-in-law of Dorothy Thorp, *née* Handel, as told to David Key, *The Supermariners* blog.

15 *Spitfire Odyssey*, p. 95.

16 *Never a Dull Moment*, p. 124

17 *Spitfire Odyssey*, p. 109.

18 *Never a Dull Moment*, p. 160.

19 Ibid.

4. Paying for the Spitfire

1 BBC Home Service, 21 May 1942.

2 *Halifax Evening Courier*, 30 August 1940.

3 'Spitfire funds: The "whip-round" that won the war?', BBC News, 12 March 2016, https://www.bbc.co.uk/news/uk-england-35697546.

4 'Spitfire: The People's Plane', BBC World Service.

5 Andrew Dennis, RAF Museum.

6 'Spitfire: The People's Plane', episode 2, BBC World Service, 17 May 2020.

7 *Leamington Spa Courier*, 23 August 1940.

8 *The Times*, 3 September 1940.

9 *Leamington Spa Courier*, 18 October 1940.

10 *Evening Despatch*, 30 August 1940.

11 'A Spitfire Wing Leader Looks Back', Duncan-Smith, quoted in *Schooldays to Spitfire*, Gordon Mitchell (History Press, 2006), p. 357.

12 Angus Archives, Facebook, https://www.facebook.com/ MontroseAirStation/posts/the-history-of-our-red-lichtie/ 2861179757293474/.
13 Andrew Dennis, RAF Museum, https://www.rafmuseum.org.uk/ blog/names-on-a-plane.
14 *The New York Times*, 22 March 1942.
15 'Spitfire funds: The "whip-round" that won the war?', ibid.
16 BBC Home Service, 24 July 1940.
17 'Military Spending', Max Roser and Mohamed Nagdy, Our World in Data, https://ourworldindata.org/military-spending.

5. The Other Spitfires

1 *Spitfire*, BBC, 6 March 1976.
2 *Spitfire: Portrait of a Legend*, Leo McKinstry (John Murray, 2007), p. 152.
3 *Spitfire Odyssey*, p. 85.
4 *Nuffield the Man*, Caroline Nixon (Nuffield Farming, 2010), p. 18.
5 Jaguar Land Rover.
6 *Never a Dull Moment*, p. 133.
7 *Spitfire*, BBC, Ibid.
8 *Spitfire, Portrait of a Legend*, p. 154.
9 *Spitfire Odyssey*, p. 85.
10 *Out on a Wing: An Autobiography*, Miles Thomas (Michael Joseph, 1964), p. 204.
11 *Never a Dull Moment*, p. 131.
12 *Spitfire*, BBC, ibid.
13 *Spitfire*, p. 154.
14 *Spitfire*, BBC, ibid.
15 *Never a Dull Moment*, p. 133.
16 *Birmingham Mail*, 1 November 2015
17 *Spitfire*, BBC, ibid.
18 *Never a Dull Moment*, ibid.
19 Ibid.
20 *Spitfire*, BBC, ibid.

6. The Inspiration of Hursley

1 Barbara Harries, daughter of Joe Smith, as told to David Key, *The Supermariners* blog.
2 *The Times*, 22 December 1942.

3 *Tomorrow is D-Day*, p. 89.

4 Ibid, p. 65.

5 Stella Rutter, *Secret Spitfires*, interview by John Beck, Beck Films, material from the Solent Sky Museum.

6 *Tomorrow is D-Day*, p. 90.

7 Stella Rutter, ibid.

8 Ibid.

7. The Father of the Spitfire

1 'The 1920s British Air Bombing Campaign in Iraq', BBC News, 6 October 2014, https://www.bbc.co.uk/news/magazine-29441383.

8. The Schneider Boost

1 *Schneider Trophy to Spitfire: The Design Career of R. J. Mitchell*, John Shelton (J. J. Haynes & Co Ltd, 2008), p. 34.

2 BBC Home Service, 18 September 1957.

3 Ibid.

4 *Schooldays to Spitfire*, Gordon Mitchell (History Press, 2006), p. 64.

5 *Flight*, 24 September 1925.

6 BBC audio archive, 1957.

7 *Schneider Trophy to Spitfire*, p. 101.

8 Ibid.

9 Ibid.

10 *The Secret Horsepower Race: Western Front Fighter Engine Development*, Calum E. Douglas (Tempest, 2020), chapter one.

11 *Schooldays to Spitfires*, p. 86.

12 *Schneider Trophy to Spitfire*, p. 136.

13 *Spitfire*, BBC, 6 March 1976.

14 *Schneider Trophy to Spitfire*, p. 136.

15 National Archives, TNA, AIR-19/128.

16 *Schneider Trophy to Spitfire*, p. 152.

17 *Adventuress: The Life and Loves of Lucy, Lady Houston*, Teresa Crompton (History Press, 2020), Kindle Edition loc 2792.

18 Jerripedia website, https://www.theislandwiki.org/index.php/Lady_Houston.

19 National Archives, TNA AIR-19/128.

20 BBC National Programme, 1 January 1931.

21 *Schooldays to Spitfires*, p. 119.

22 *Spitfire*, BBC, ibid.
23 *The Secret Horsepower Race*, p. 25.

9. A New Fighter Needed

1 BBC News website, 6 October 2014, https://www.bbc.co.uk/news/magazine-29441383.
2 *Forty Years of the Spitfire,* Jeffrey Quill.
3 Robin Baker, son of Hazel, interview with the author.
4 *The Spitfire Pocket Manual,* Martin Robson (Osprey, 2010).
5 *Spitfire,* Leo McKinstry, p. 53.
6 *The Schoolgirl Who Helped to Win a War*, BBC, 11 July 2020.
7 Robin Baker, son of Hazel, interview with the author.

10. The Spitfire Spy

1 Farnborough Air Sciences Trust, Briefing 9.
2 Ibid.
3 National Archives, KV-2-2200-1.
4 National Archives, KV-2-2200-2.

11. Meanwhile in Germany

1 Carl von Ossietzky, Nobel Prize website, https://www.nobelprize.org/prizes/peace/1935/ossietzky/biographical.
2 'Supermarine S6 and S6B', BAE Systems, https://www.baesystems.com/en//en/heritage/supermarine-s6.
3 *The Secret Horsepower Race,* p. 37.
4 Ibid, p. 40.
5 Ibid, p. 41.
6 Ibid, p. 441.

12. Data and Danger

1 *Spitfire*, BBC, 6 March 1976.
2 Ibid.
3 *Spitfire,* Leo McKinstry, p. 44.
4 The *Twentieth Century House in Britain: From the Archives of Country Life,* Alan Powers Aurum Press, 2004.
5 Jeffrey Quill obituary, *Guardian*, 5 March 1996.
6 *Test Pilot*, BBC, Tuesday Documentary, 1 September 1970.

7   *The Battle of Britain*, Richard Hough and Denise Richards, (Pen & Sword Aviation: 2007), p.40.

8   *Test Pilot*, BBC, ibid

9   *Spitfire*, BBC, ibid.

10 IWM, 10687

11 Ibid.

12 Ibid.

13 *A Short History of Aviation Gasoline Development 1903–1980*, Alexander R. Oyston, 'The Aeronautical Journal', December 1981.

14 *The Secret Horsepower Race*, p. 119.

15 *Test Pilot*, BBC, ibid

16 *Spitfire: A Test Pilot's Story*, p. 168.

17 *Spitfire Odyssey*, p, 161.

18 *Spitfire: A Very British Love Story*, John Nichol (Simon & Schuster, 2018), p. 80.

19 IWM, ibid.

20 Ibid.

13. Miss Shilling's Orifice

1   *Spitfire*, BBC, 6 March 1976.

2   *The Secret Horsepower Race*, p. 174.

3   *Spitfire: The People's Plane*, episode 6, Donald Spiers, 14 June 2020.

4   *Spitfire*, BBC, ibid.

5   'Embarking on an Engineering Career in the Twenties', Beatrice Shilling, *The Woman Engineer*, volume 10-1969, courtesy of the Women's Engineering Society and the IET Archives Women's Engineering Society.

6   *Negative Gravity: A Life of Beatrice Shilling*, Matthew Freudenberg (Charlton Publications, 2003), p. 14.

7   *The Woman Engineer*, ibid.

8   *Spitfire: The People's Plane*, episode 6, Jo Denbury, 14 June 2020.

9   'Embarking on an Engineering Career in the Twenties', Beatrice Shilling.

10 *Negative Gravity*, p. 25.

11 *The Woman Engineer*, ibid.

12 *Negative Gravity*, p. 59

13 Ibid, p. 42.

14 *The Secret Horsepower Race*, p. 174.

15 *Spitfire*, BBC, ibid.

Notes

16 *Spitfire: The People's Plane*, episode 6, Jo Denbury, 14 June 2020.
17 *The Woman Engineer*, ibid.
18 *Spitfire: A Test Pilot's Story*, p. 233; John Peel obituary, *Daily Telegraph*, 16 January 2004.

14. The Spitfire Becomes the Star

1 *Battle of Britain On Screen: The Few in British Film and Television Drama*, S. P. Mackenzie (Bloomsbury, 2007), p. 24.
2 Ibid, p. 25.
3 Ibid, p. 31.
4 *Time* magazine, 14 June 1943.
5 *The New York Times*, 3 June 1943.
6 'Why Did the Nazis Murder Leslie Howard?', *The Lady*, https://lady. co.uk/why-did-nazis-murder-leslie-howard.

15. A Deadly New Foe

1 *Wing Leader*, J. E. 'Johnnie' Johnson (Chatto and Windus, 1956).
2 *Spitfire: A Test Pilot's Story*, p. 206.
3 *Spitfire*, BBC, 6 March 1976.
4 IWM, 10687.
5 *Spitfire: A Test Pilot's Story*, p. 210.
6 *The Secret Horsepower Race*, p. 239.
7 Ibid, p. 240.
8 IWM, ibid.

16. Hursley Matches the Germans

1 *Tomorrow Is D-Day*, p. 124.
2 Ibid.
3 American Heritage Museum website – https://www. americanheritagemuseum.org/aircrafts/supermarine-spitfire-mk-ix/ #:~:text=%E2%80%9CThe%20performance%20of%20the%20 Spitfire,and%20climb%20is%20exceptionally%20good.
4 Stella Rutter, interview with John Beck.

17. A Bullet Through the Sky

1 Sir Keith Park website, sirkeithpark.com.
2 *Ramblings*, BBC Radio Four, 14 November 2008.

3 *Midweek*, BBC Radio Four, 24 April 2002.

4 *First Light*, Geoffrey Wellum (Penguin, 2002), p. 1.

5 *Spitfire!*, Palmer and Whitehead, BBC4, 26 September 2019.

6 Ibid.

7 Ibid.

8 Ibid.

9 *Battle of Britain*, BBC1, 19 September 2010.

10 *Spitfire!*, BBC4, ibid.

11 *Battle of Britain*, BBC1, ibid.

12 *Midweek*, BBC Radio Four, ibid.

13 *First Light*, p. 153.

14 *First Light*, Whiteman, BBC2, 14 September 2010.

15 *Spitfire*, BBC Radio Drama, 9 November 2010.

16 *Battle of Britain*, BBC1, ibid.

17 *Midweek*, BBC Radio Four, ibid.

18 *First Light*, BBC2, ibid.

19 *Battle of Britain*, BBC1, ibid.

20 *First Light*, BBC2, ibid.

21 *Ramblings*, BBC Radio Four, ibid.

22 Ibid.

23 *First Light*, p. 337.

18. Women Take the Controls

1 IWM, 8659.

2 IWM, 9359.

3 Ibid, 8659.

4 Ibid.

5 'In the Soup', David Trotter, *London Review of Books*, 9 October 2014, *https://www.lrb.co.uk/the-paper/v36/n19/david-trotter/in-the-soup*.

6 IWM, ibid.

7 *Ferry Pilots*, BBC Home Service, 20 February 1942.

8 IWM, ibid.

9 IWM, 9359.

10 Ibid, 8659.

11 Giles Whittell, in *Spitfire: The People's Plane*, episode 5, BBC World Service, 7 June 2020.

12 IWM, 8659.

13 Giles Whittell, in *Spitfire: The People's Plane*, ibid.

14 IWM, 8659.

15 Ibid, 9579

16 Ibid.

17 *Spitfire Women*, BBC Radio Wales, 21 September 2008.

18 *Spitfire Women of World War II*, Giles Whittell (Harper Press, 2008).

19 *Spitfire Women*, BBC Radio Wales, ibid.

20 Ibid.

21 *Irish Times*, 2 July 2003.

22 *Pioneers of Aviation*, Southjets.com, https://southjets.com/en/pioneers-of-aviation.

23 *Miss Frost*, BBC Radio Wales, 12 December 2015.

24 Ibid.

25 *Spitfire Women*, BBC Radio Wales, ibid.

26 IWM, 9362.

27 Ibid, 9579.

28 Pauline Gower, quoted in *A Harvest of Memories: The Life of Pauline Gower MBE*, Michael Fahie (GMS Enterprises, 1995).

29 Ibid, p. 170.

30 *Spitfire Women*, BBC Radio Wales, ibid.

19. The Yanks are Coming

1 *Tally Ho! Yankee in a Spitfire*, Art Donahue (independently republished, 2014), chapter four, loc. 659, references from Kindle edition.

2 *Democrat and Chronicle*, 6 August 1940.

3 Ibid.

4 *American Eagles: American Volunteers in the RAF 1937–43*, Tony Holmes (Midland, 2001), p. 102.

5 *Tally Ho*, loc. 60.

6 'Colonel Reade Tilley', *PM*, BBC Radio Four, 2 September 1976.

7 *War Eagles*, Colonel James Saxon Childers (Tannenberg, 2016), loc. 395.

8 *Tally Ho*, loc. 928.

9 Ibid, loc. 968.

10 Ibid, loc. 1036.

11 Ibid, loc. 1151.

12 Ibid, loc. 1151/1178.

13 Ibid, loc. 2221.

14 Ibid, loc. 1965.

15 'The Untold Story of the RAF's Black Second World War Fliers Over Europe', Mark Johnson, National Archives podcast, 11 June 2014.

16 'Answering the Call', RAF Museum, https://www.rafmuseum.org.uk/research/online-exhibitions/pilots-of-the-caribbean/answering-the-call/the-second-world-war-1939-to-1945-recruitment.aspx.

17 'The Untold Story of the RAF's Black Second World War Fliers Over Europe', ibid.

18 BBC Home Service, 29 September 1942.

19 American Air Museum, http://www.americanairmuseum.com/person/244715.

20 'Colonel Reade Tilley', *PM*, Ibid.

21 American Air Museum, http://www.americanairmuseum.com/person/12239.

20. Plotting the Future

1 Diary of Peggy Balfour, held by Imperial War Museum.

2 Ibid.

3 Ibid.

4 Ibid.

5 IWM sound archive 11544.

6 Diary of Peggy Balfour, ibid.

7 Ibid.

8 Ibid.

9 Ibid.

10 Jack Lawson combat report, courtesy of IWM Duxford.

11 Diary of Peggy Balfour, ibid.

12 *A Fine Blue Day*, BBC Radio Four, 1 August 1978.

13 BBC Home Service, 1 January 1961.

14 *Life* magazine, 28 February 1949, p. 48.

15 IWM, 2807.

16 Ibid.

17 Ibid.

18 Jack Lawson combat report, ibid.

19 battleofbritain1940.net.

20 *Battle of Britain: A Day-to-Day Chronicle: 10 July–31 October 1940*, Patrick Bishop (Quercus, 2020), p. 307.

21 Diary of Peggy Balfour, ibid.

22 *Persuading the People: British Propaganda in World War II*, David Welch (British Library 2016) p. 131.

Index

Index